Handling Qualities Technical Meeting 2025

Huntsville, Alabama, USA
22 - 23 January 2025

ISBN: 979-8-3313-2655-5

Printed from e-media with permission by:

Curran Associates, Inc.
57 Morehouse Lane
Red Hook, NY 12571

Some format issues inherent in the e-media version may also appear in this print version.

Vertical Flight Society
2700 Prosperity Ave, Suite 275
Fairfax, VA 22031
USA

Phone: (703) 684-6777
Fax: (703) 739-9279

www.vtol.org

Additional copies of this publication are available from:

Curran Associates, Inc.
57 Morehouse Lane
Red Hook, NY 12571 USA
Phone: 845-758-0400
Fax: 845-758-2633
Email: curran@proceedings.com
Web: www.proceedings.com

VFS Handling Qualities Meeting Agenda
January 22-23, 2025

Wednesday, January 22, 2025	
0730-1200	***Attendee Registration***
0730-1200	***Presentation Material Drop-off for Authors***
0800-0815	**Welcome by Chapter President –** Dr. David O'Brien
	Meeting Objectives by General Chair – Mr. Scott Smith
0815 - 0900	**Opening Keynote: COL Ryan C. Nesrsta,** **US Army, Program Manager, Utility Helicopters** *Path Forward for UHPO & Thoughts on Autonomy for the Enduring Fleet and FVL*
	Session I: Test and Evaluation
0900-0930	*A Comprehensive Overview of the Handling Qualities Studies Conducted in the Vertical Motion Simulator for the Advanced Rotorcraft Configuration Testing of Increased Capabilities...1* Tom Berger, Carl Ott, U.S. Army Combat Capabilities Development Command, AvMC
0930-1000	*Flight Loads and Aeromechanics Characterization of a Soft Inplane Hingeless Rotor Control System during Handling Qualities Testing...20* Marc Alexander, National Research Council of Canada
1000-1015	**Break**
1015-1045	*Preliminary Evaluation of Stability and Control of a Lift + Cruise eVTOL Configuration...34* Justin Martin, U.S. Army Combat Capabilities Development Command, AvMC
1045-1115	*Preliminary Findings from FAA sponsored Flight Tests of Handling Qualities Task Elements for VTOL Aircraft...47* Martin Shubert, Tiltrotor Flight Test Consulting LLC, Mike Jones, Systems Tech. Inc.
1115-1145	*Airbus H145...58* Carl Forsling, Airbus
1145-1245	**Lunch**
1245-1345	**Day 1 Keynote: SES Dr. Stephanie Reitmeier** **Director of the Software, Simulation, Systems Eng. Integration** **Directorate (S3I), US Army DEVCOM AvMC**
	Session II – Flight Controls
1345-1415	*Evaluation of Active Sidestick Failure Mode Handling Qualities: Paper...66* Tom Berger, U.S. Army Combat Capabilities Development Command, AvMC; Kevin Christensen, Seung Baik & John (Jack) Rominger, Bell Textron Inc., Sam Clark, IronMountain Solutions
1415-1430	**Break**
1430-1500	*An Analytical Study of Helicopter Handling Qualities Characteristics due to Active Force-Sensing Side-Stick Utilization...79* Marc Alexander, National Research Council of Canada
1500-1530	*Capital Architecture...95* Lyle Chamberlin, Near Earth Autonomy
1530-1600	*Handling Qualities for the Valiant: Presentation Not Available for Distribution* Larry French, Griffon Aerospace
1600-1800	**Social**

Thursday, January 23, 2025	
0730-1200	**Attendee Registration**
0730-1200	**Presentation Material Drop-off for Authors**
0830-0845	**Day 2 Welcome by Technical Chair –** Mr. Eric Grigorian
	Session III: Human Factors
0845-0915	*Human Factors System Safety Processes for Flight Director Modes...103* Tia Larsen-Calcano, PM FLRAA
0915-0945	*Apache Pilot Workload: Discussion Only (No Presentation)* COL Daniel Thetford, US Army, Program Manager, Apache Helicopter
0945-1015	*Handling Qualities MTEs Evaluation Methodologies in Preliminary Design of High-Speed Rotorcraft Configurations...112* Fabio Riccardi, Matteo Mamino, Matteo Pecoraro, Andrea Ragazzi, Leonardo Helicopters
1015-1030	**Break**
1030-1100	*PIO and Handling Qualities Prediction Using the USAFTPS Bjorkman PIO Data Set...124* Ed Bachelder, San Jose State University; Eileen Bjorkman, Air Force Test Center; Bimal Aponso, Federal Aviation Administration
1100-1130	*Usability Shortcomings in Department of Defense Funded Systems...146* Jerry Burpee, Intuitive Research
1130-1200	*Rotor Blade Position Simulation from Dynamic Maneuvers for Launched Effects...153* Chris Brackbill, U.S. Army Combat Capabilities Development Command, AvMC
1200-1300	**Lunch**
1300-1345	**Day 2 Keynote: BG David C. Phillips,** **US Army, Program Executive Officer, Aviation**
1400-1600	**Pilot Roundtable: BOGSAT** with Carl Ott, Alan Davis, Rodney Allison, BG David Phillips, COL Minor
1600 - 1615	**Closing –** President, Redstone Chapter

Vertical Flight Society

APPROVED FOR PUBLIC RELEASE

U.S. ARMY

DEVCOM AVIATION & MISSILE CENTER

ADVANCED ROTORCRAFT CONFIGURATIONS TESTING OF INCREASED CAPABILITIES (ARCTIC)

Presented at: 2025 VFS Handling Qualities Technical Meeting | 22 JAN 2025

TOM BERGER, DEVCOM AVMC, MOFFETT FIELD, CA
CARL OTT, DEVCOM AVMC, FT EUSTIS, VA

APPROVED FOR PUBLIC RELEASE

OUTLINE

DEVCOM AVIATION & MISSILE CENTER

- ARCTIC Experimentation Background
- Accomplishments
- Enablers
- Summary
- What's Next

APPROVED FOR PUBLIC RELEASE

ARCTIC EXPERIMENTS BACKGROUND

DEVCOM AVIATION & MISSILE CENTER

- **Evolved from capabilities developed to study agility of advanced configurations for Army's Operational Concepts Team (OCT)**
- **Objective: provide enduring, agile, and independent Government FVL simulation environment for use by Program Managers and S&T SMEs to:**
  - Expose Army aviators to FVL representative configurations and technologies
  - Conduct design trade studies and inform requirements
  - Refine handling qualities requirements for FVL advanced configurations
  - Explore increased control system capabilities and failure modes
- **Leverage generic flight-dynamics models, control laws developed by U.S. Army Combat Capabilities Development Command Aviation & Missile Center (DEVCOM AvMC)**

AvMC Generic FLRAA-Size Models

AvMC Generic FARA-Size Models

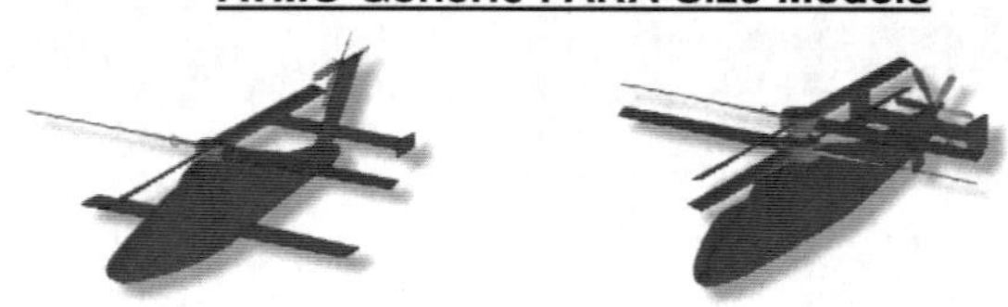

ARCTIC TIMELINE

DEVCOM AVIATION & MISSILE CENTER

2021 › 2022 › 2023 › 2024 › 2025

ARCTIC 1 — OCT-NOV 2021
- FLRAA candidate configuration HQ study
- TCL vs collective

ARCTIC 2 — APR-JUN 2022
- FARA HQ study
- Linked / unlinked inceptor study
- FLRAA coax configuration damage tolerant control study

ARCTIC 3 — FEB-MAR 2023
- FARA yaw axis study
- High-speed MTE development
- FLRAA tiltrotor configuration NVG terrain flight study

ARCTIC 4 — SEP 2023
- FARA workload study

ARCTIC 5 — AUG 2024
- FLRAA tiltrotor configuration terrain flight study
- Tiltrotor sidestick failure mode study

ARCTIC 6 — Spring 2025
- FLRAA tiltrotor configuration autonomy study

Accomplishments

KEY ARCTIC ACCOMPLISHMENTS

- ✓ Demonstrated feasibility / benefit of thrust control lever for advanced FVL configurations
- ✓ Demonstrated impact of linked vs. unlinked controls on crew coordination
- ✓ Tested proposed high-speed MTEs in motion-based simulator
- ✓ Developed new terrain-flight surrogate high-speed MTEs
- ✓ Proposed updates to yaw-axis target acquisition & tracking requirement boundaries
- ✓ Assessed handling qualities and survivability benefits of damage tolerant control
- ✓ Investigated novel response types
- ✓ Quantified workload and mission effectiveness benefits of full-authority fly-by-wire
- ✓ Assessed ability to fly high-speed terrain flight with NVG
- ✓ Quantified impacts of airspeed on workload during tiltrotor terrain flight
- ✓ Published eight technical papers

DEVCOM AVIATION & MISSILE CENTER

Enablers

APPROVED FOR PUBLIC RELEASE

APPROVED FOR PUBLIC RELEASE

KEY ARCTIC ENABLERS

DEVCOM AVIATION & MISSILE CENTER

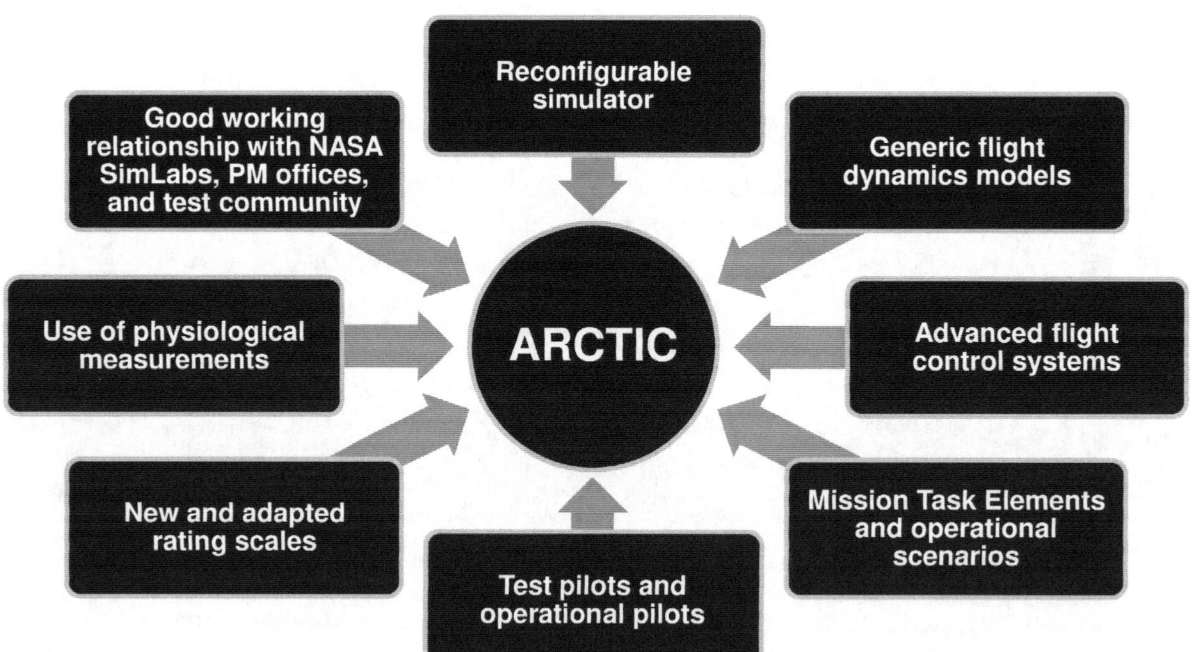

APPROVED FOR PUBLIC RELEASE

KEY ARCTIC ENABLERS

DEVCOM AVIATION & MISSILE CENTER

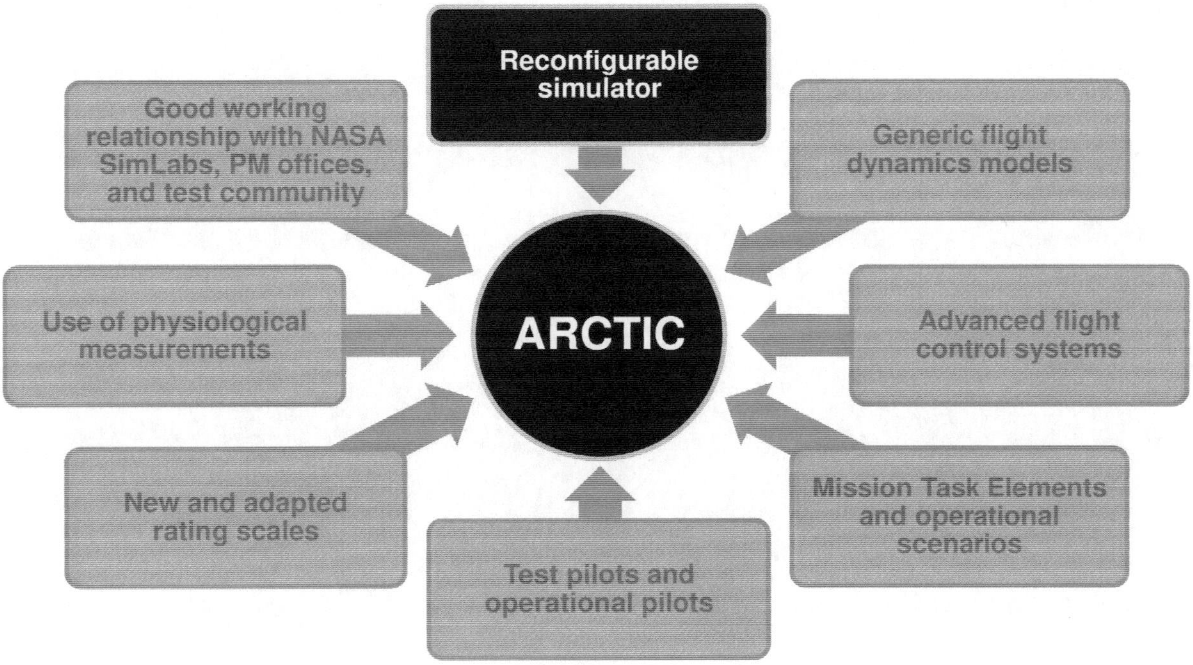

NASA VERTICAL MOTION SIMULATOR (VMS)

DEVCOM AVIATION & MISSILE CENTER

- **Large displacement six independent DOF**
  - ±30 ft vertical, ±20 ft lateral, ±4 ft longitudinal
  - 0.7 g vertical acceleration
- **Configurable side-by-side rotorcraft cockpit (T-cab)**
  - Collective and thrust control lever (TCL)
  - Linked and unlinked inceptors
  - Cockpit divider to emulate tandem cockpit
  - Panel mounted displays based on CAAS and ICE
  - Configurable audio
- **DVE simulations**
  - Night vision googles (projectors in IR mode)
  - Dynamic environmental conditions (fog, dust)
- **Integration of third-party software / hardware**
  - AFSIM for real-time threat modeling
  - Physiological sensors

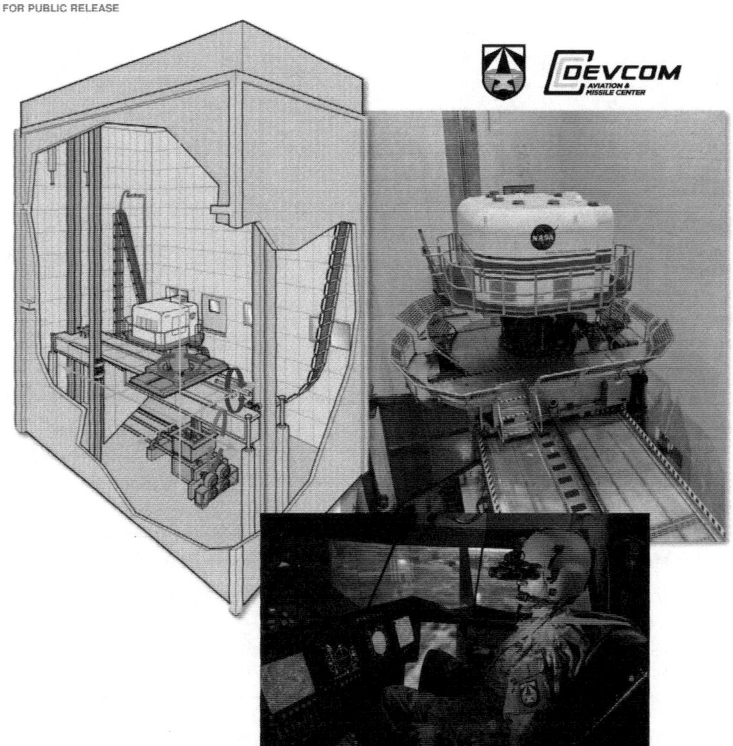

Army Aviation Magazine, Dec. 2024

KEY ARCTIC ENABLERS

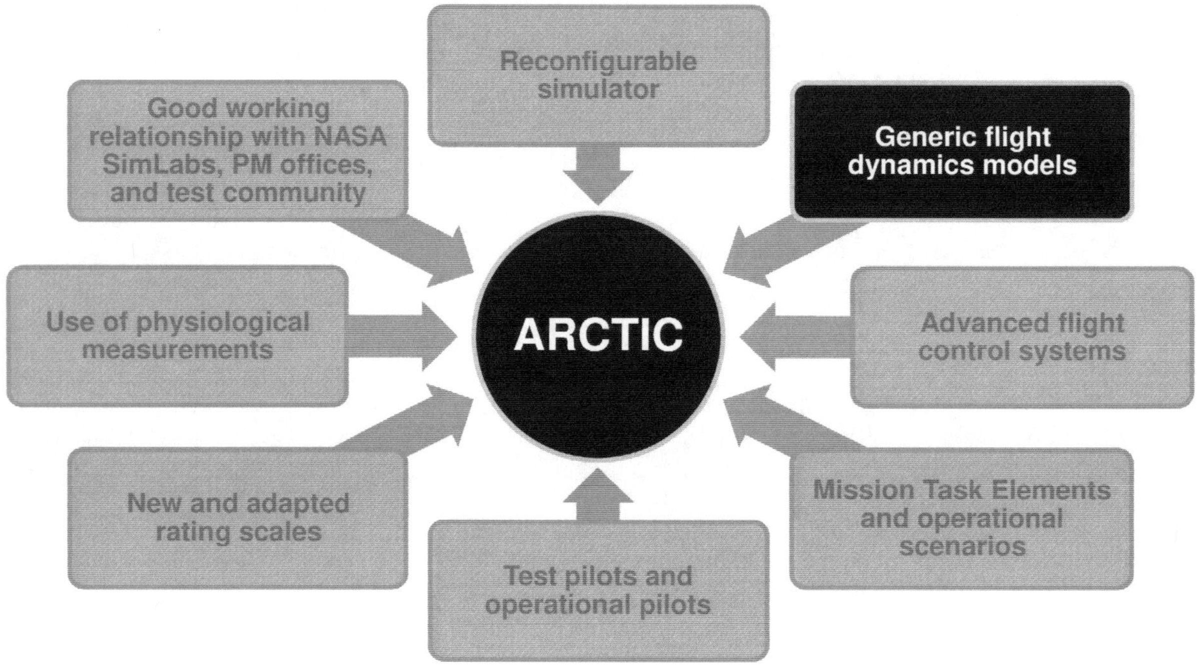

11

MODELS

- **High-fidelity flight dynamics models representative of FVL configurations**
  - Generic in nature
  - Open-source data
- **Modeled using comprehensive rotorcraft simulation code HeliUM-A**
  - Blade element, flexible blade model
  - Extensive validation against flight data
- **Real-time simulation models based on stitched model architecture**
  - Developed from high-order (~50 state) linear models and trim data

No OEM data or data rights restrictions

AvMC Generic FLRAA-Size Models

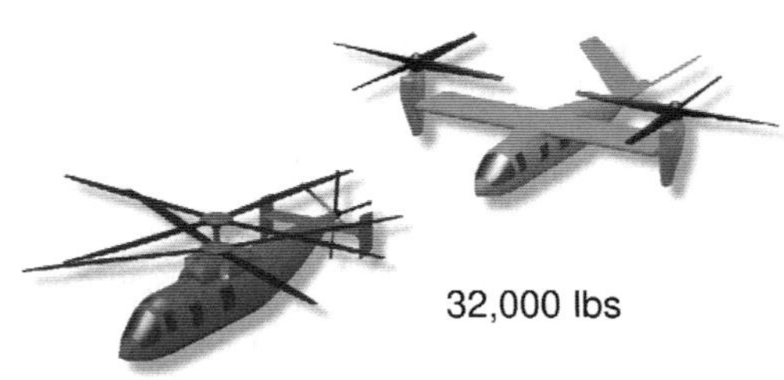

AvMC Generic FARA-Size Models

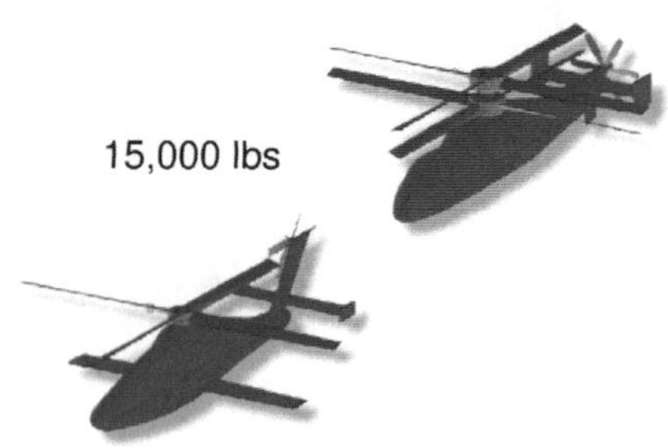

12

KEY ARCTIC ENABLERS

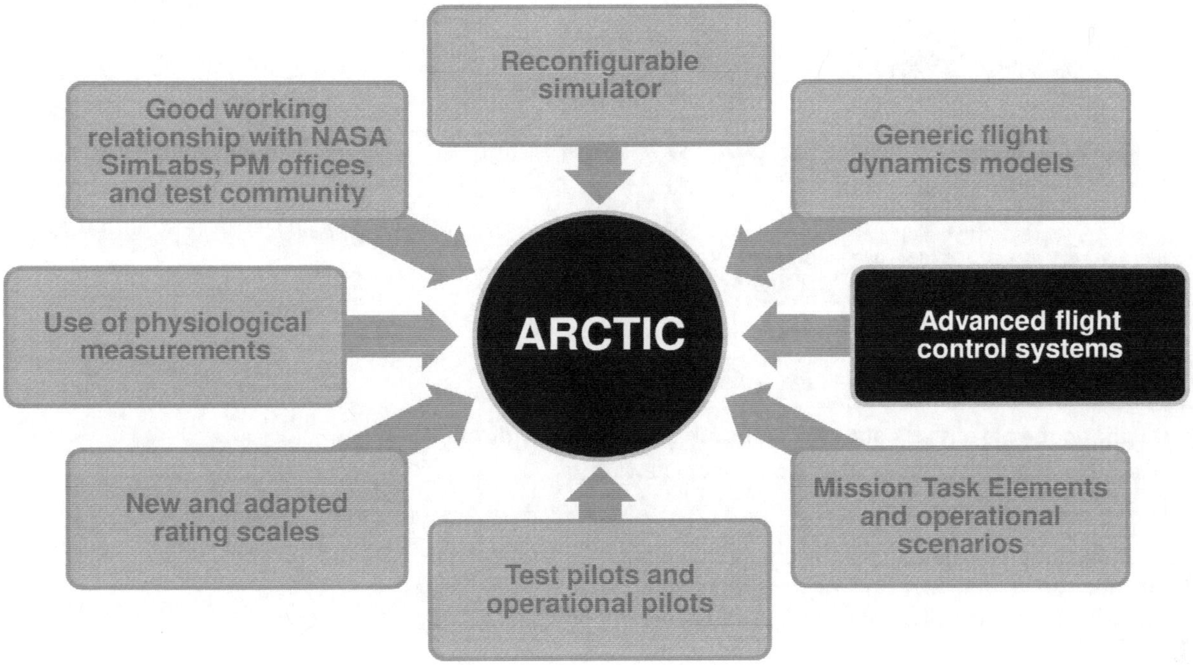

CONTROL LAWS: OVERVIEW

- **Government developed control laws for all configurations**
  - Optimized in CONDUIT to meet Level 1 MIL-DTL-32742 requirements
- **Explicit model following architecture provides flexibility**
  - Easily change response types
  - Pilot selectable levels of augmentation
- **Weighted pseudo inverse control allocation**
  - Sends commands to rotor and aerosurface actuators as function of airspeed and nacelle angle

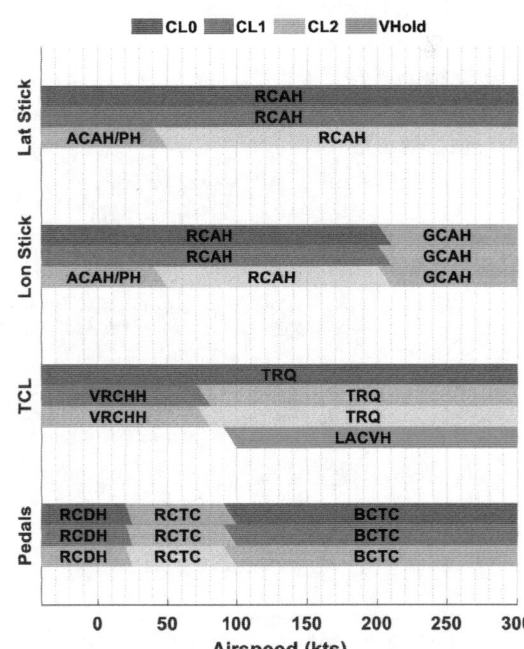

CONTROL LAWS: AUTO MODE

DEVCOM AVIATION & MISSILE CENTER

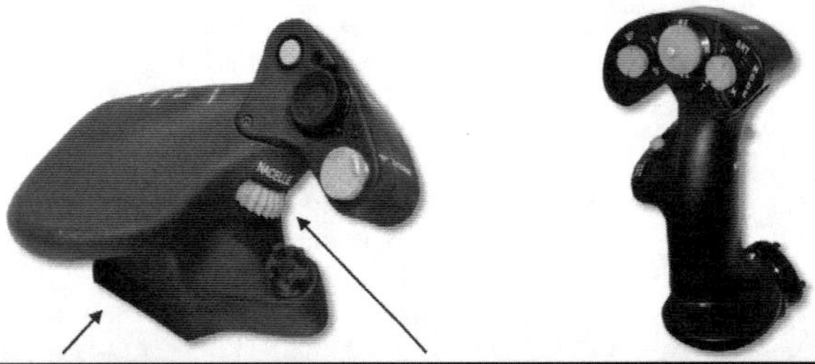

	TCL	Thumb Wheel	Sidestick
Baseline	Power	Nacelle / prop	Pitch attitude
Auto Mode	Faster / slower	--	Up / down

- **Auto Mode: pilot controls aircraft states (flight path and airspeed) directly**
  - Similar inceptor configuration to F-35B Unified Control Laws
  - Automatic control of prop / nacelle angle

Improved transition and high-speed HQ; ready for integration with outer guidance / autonomy loops

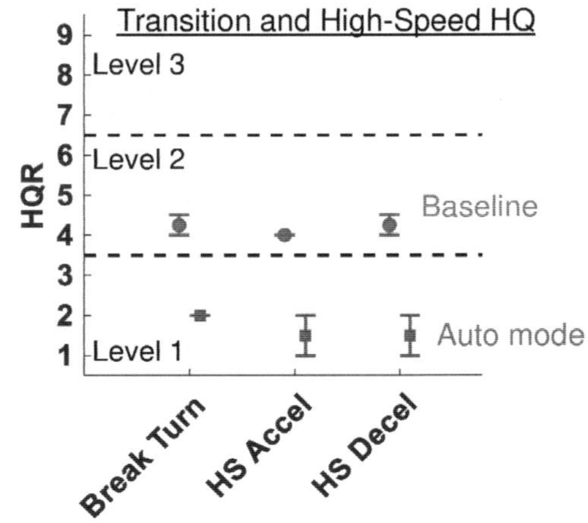

CONTROL LAWS: UPDATES FOR DAMAGE TOLERANT CONTROL (DTC)

DEVCOM AVIATION & MISSILE CENTER

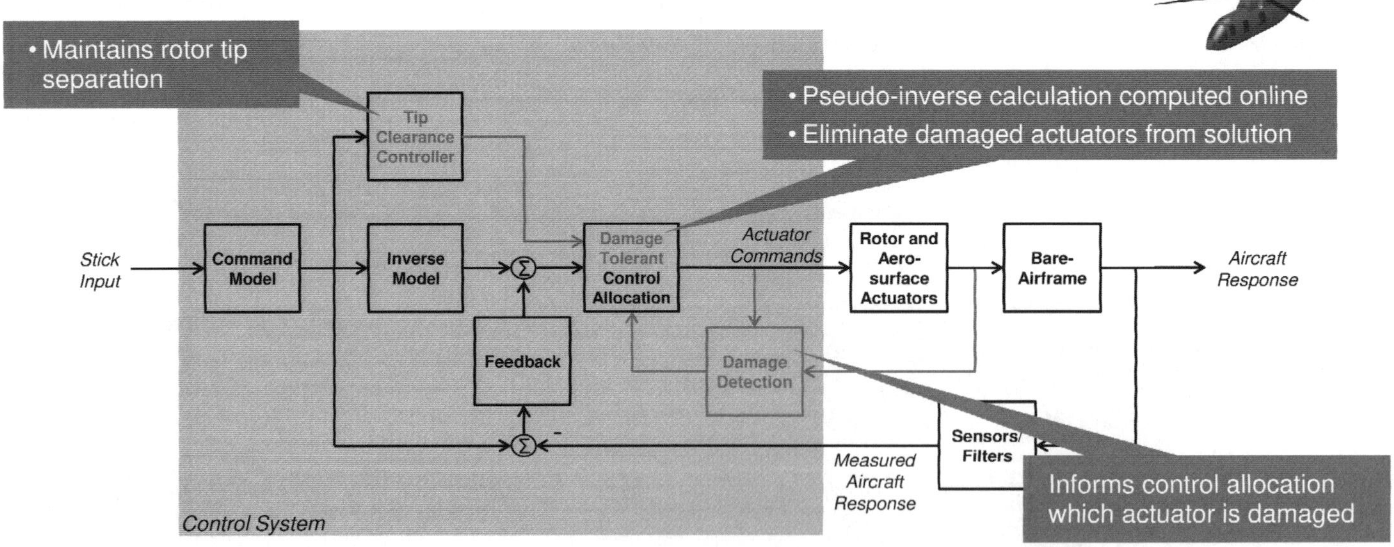

Significant reduction in control coupling and workload, and improved tip clearance with DTC

KEY ARCTIC ENABLERS

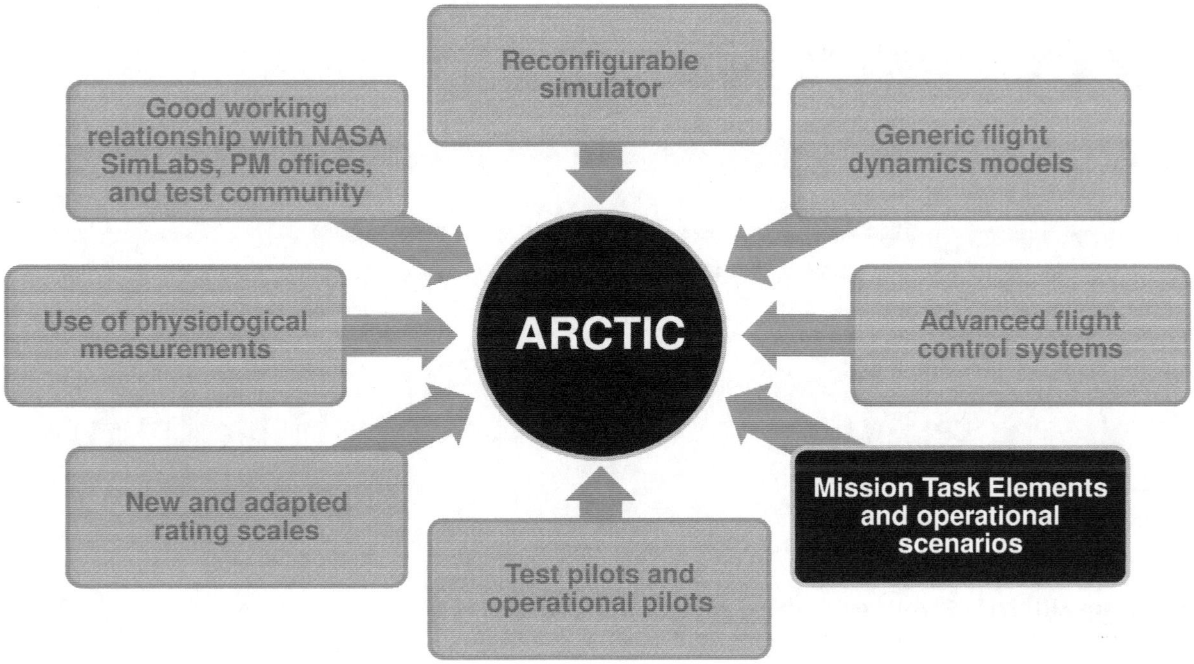

MISSION TASK ELEMENTS

- **New MTEs developed to support required testing**
  - Example: surrogate high-speed terrain flight tasks developed with DLR

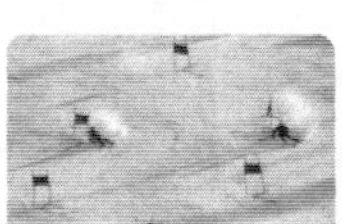

Giant Slalom
Lateral Flight Path Change

Big Air
Vertical Flight Path Change

Super Combined
Vertical & Lateral Flight Path Change

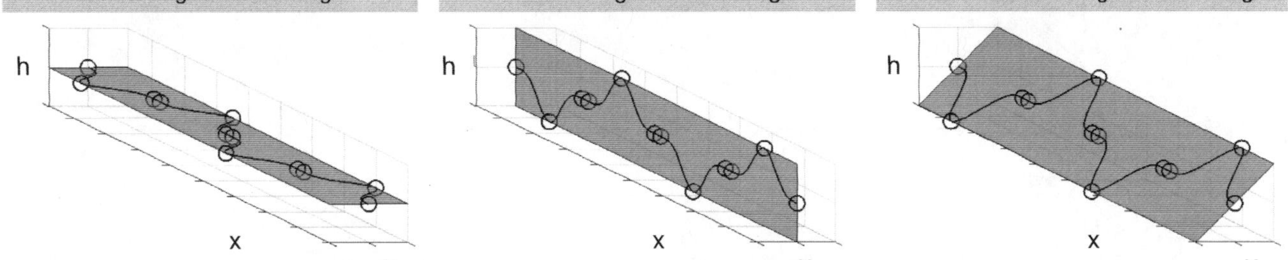

Able to vet new MTEs in simulator before going to flight
Plan to test these MTEs in flight with augmented reality head mounted display

DLR has authorized public release of surrogate high-speed terrain flight tasks developed with DLR

MISSION TASK ELEMENTS

- **New and adapted yaw-axis MTEs developed to assess location of yaw-axis Target Acquisition & Tracking requirement boundaries**

Hover Target Acquisition

(Adapted from 1992 VMS Study)

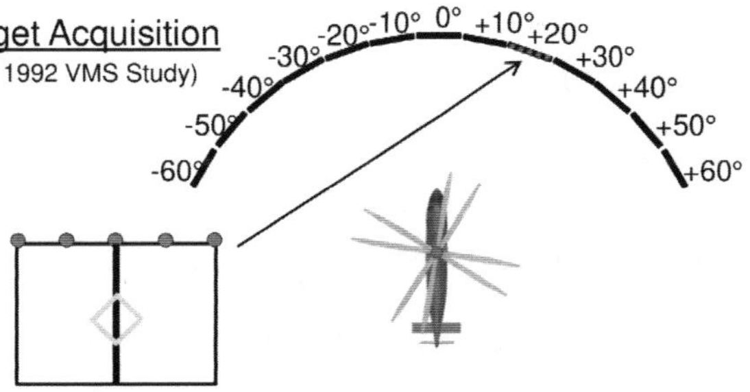

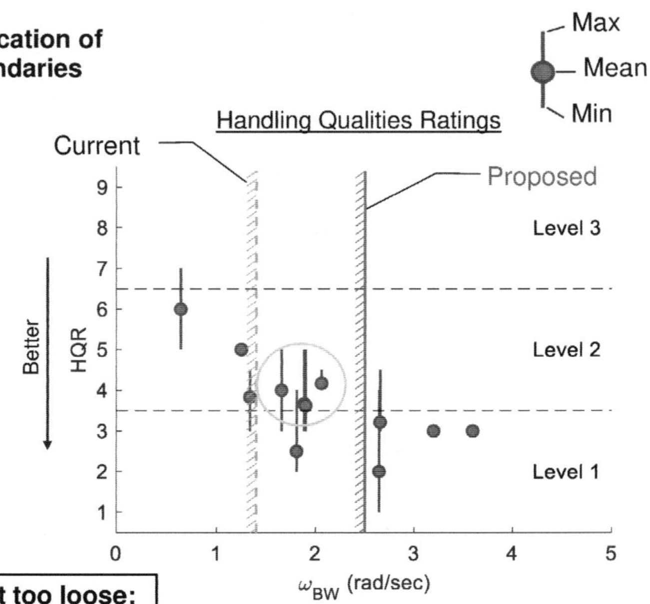

Showed current MIL-DTL-32742 yaw bandwidth requirement too loose; proposed boundary better predicts Level 1 and Level 2 designs

OPERATIONAL SCENARIOS

- **Operational scenarios (mission vignettes) developed to supplement Mission Task Elements testing**
  - Exercise full flight envelope (transition/conversion)
  - Threat models using AFSIM

Terminal Area Recovery & Rescue Operational Scenario

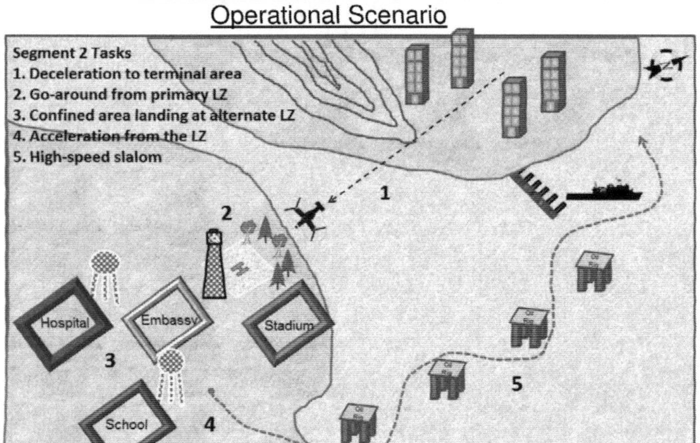

Terrain Flight Scenario

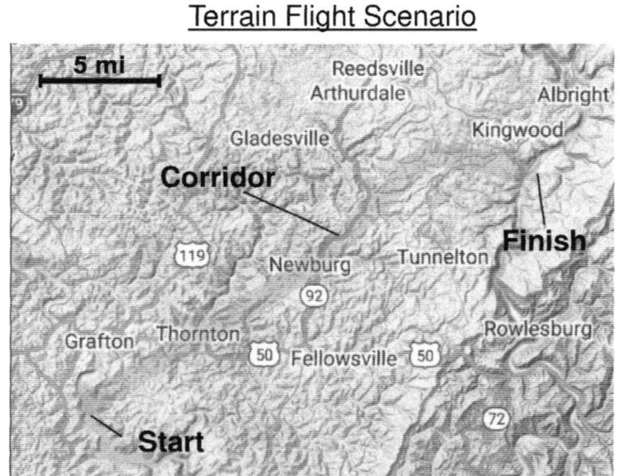

Note: scenarios for simulation purposes only and do not represent an actual Army operational scenario

OPERATIONAL SCENARIOS EXAMPLE

- **Operational scenarios developed for linked vs. unlinked sidestick study**
  - Designed to induce transfer of controls between pilots
- **Events injected to induce control transfer**
  - Imminent bird strike or midair collision
  - Brownout / fog that only affected one pilot
  - Hostile fire
  - NVG failure
- **Realistic distractions**
  - ATC / ground force communication, warnings / cautions

> **"Scenarios made it easier to tease out differences, made linked preference more noticeable."**

Example Scenario

Note: scenario for simulation purposes only and does not represent an actual Army operational scenario

KEY ARCTIC ENABLERS

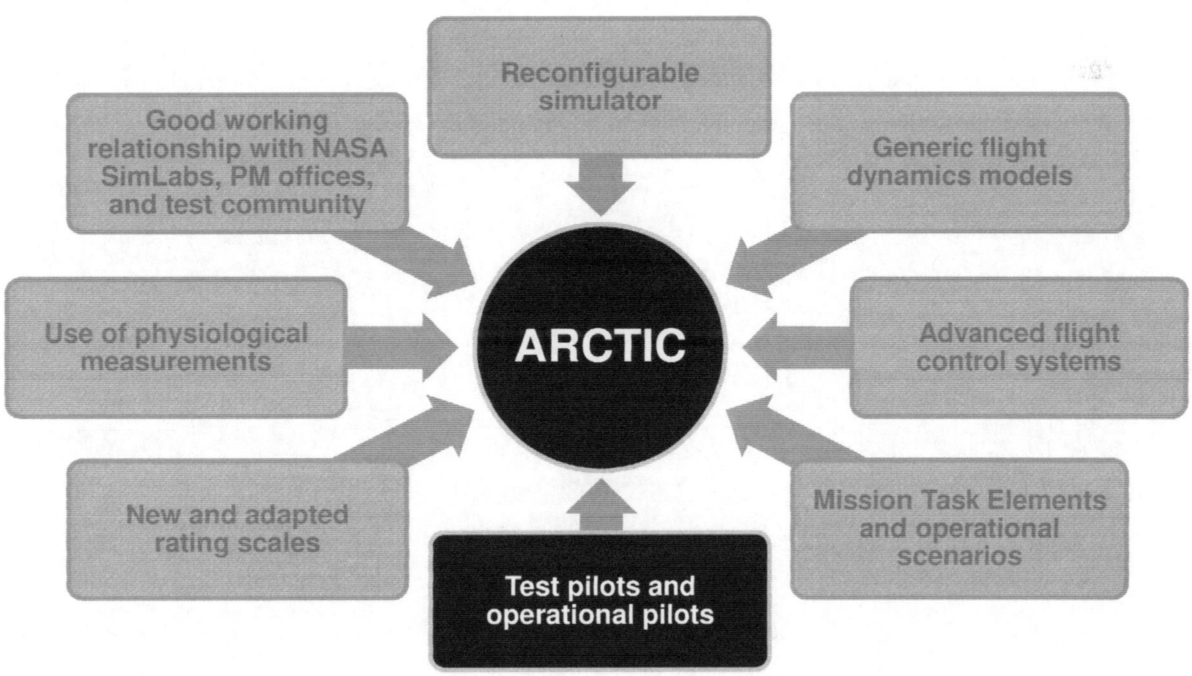

PILOTS

- **83 pilots across 5 sim events**
- **Combination of experimental test pilots (XPs), operational pilots (OPs), and industry pilots**
  - Including pilots from requirements and training / doctrine organizations
- **Range of experience**
  - Hours: 250 – 8000+
  - Aircraft: H-60A/L/M, H-47F/G, AH-64D/E, OH-58D, V-22, C-12

Army Aviation Magazine, Dec. 2024

Beneficial to keep operational users tied in with S&T efforts to ensure operationally relevant feedback

KEY ARCTIC ENABLERS

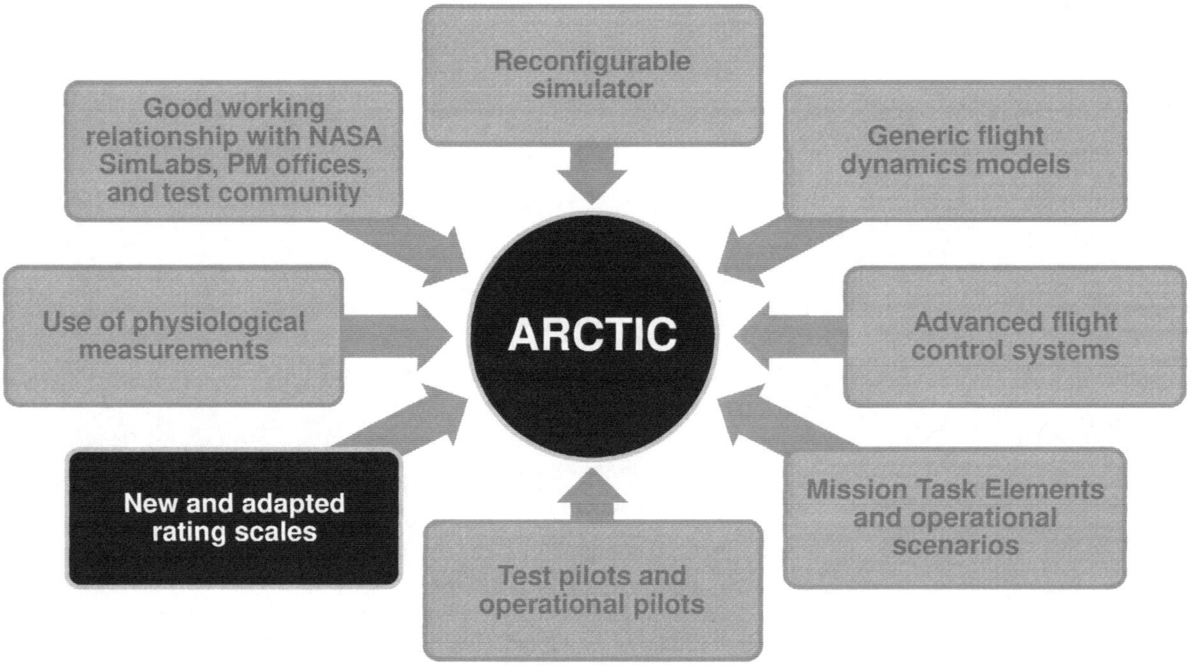

NEW RATING SCALES

- **New and adapted rating and assessment methods**
  - Agility scale
  - Predictability scale
  - Ride quality scale
  - Acceptance scale
  - Mission effectiveness scale
  - Survivability scale
  - Estimated fatigue

Mission Effectiveness Scale

#	Effectiveness	Description
1	Highly Effective	Able to complete mission with little to no excursions outside of task parameters.
2	Mostly Effective	Some infrequent or large excursions outside of task parameters during mission.
3	Somewhat Effective	Mission completion possible with frequent excursions outside parameters.
4	Marginal Effectiveness	Some mission tasks were shed to control aircraft.
5	Ineffective	Unable to complete mission; control of aircraft requires full attention of pilot.

Acceptance Scale

Item	5-Point Scale						
1	Helpful	1	2	3	4	5	Dangerous
2	Natural	1	2	3	4	5	Unnatural
3	Bad	1	2	3	4	5	Good
4	Intuitive	1	2	3	4	5	Confusing
5	Effective	1	2	3	4	5	Superfluous
6	Irritating	1	2	3	4	5	Likeable
7	Decrease Workload	1	2	3	4	5	Increase Workload
8	Undesirable	1	2	3	4	5	Desirable
9	Increase Awareness	1	2	3	4	5	Distracting

25

NEW RATING SCALES: EXAMPLE RESULTS

Predictability & Awareness Ratings

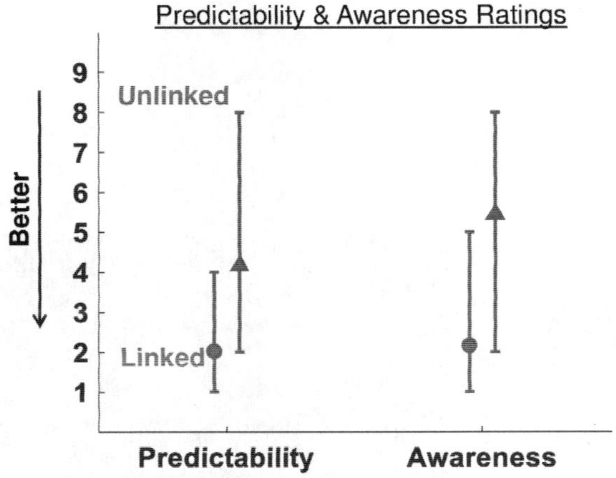

Acceptance Scale Ratings

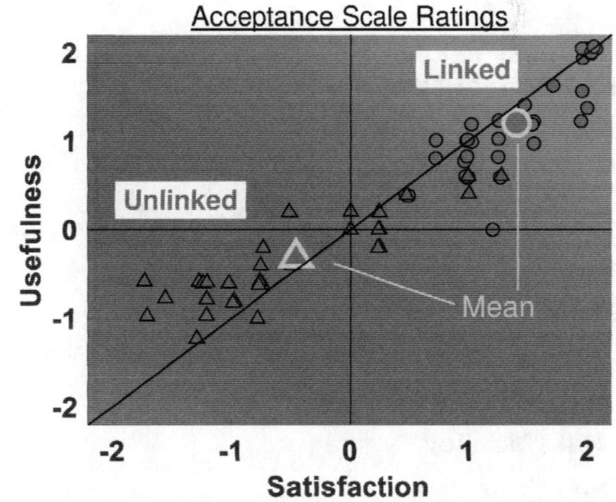

- Linked sticks showed marked improvement in both awareness and predictability rating
- Acceptance scores solidly in positive quadrant for Linked, in negative quadrant for Unlinked

Unlinked sticks reduced pilots' awareness and ability to predict aircraft response when not in control

26

KEY ARCTIC ENABLERS

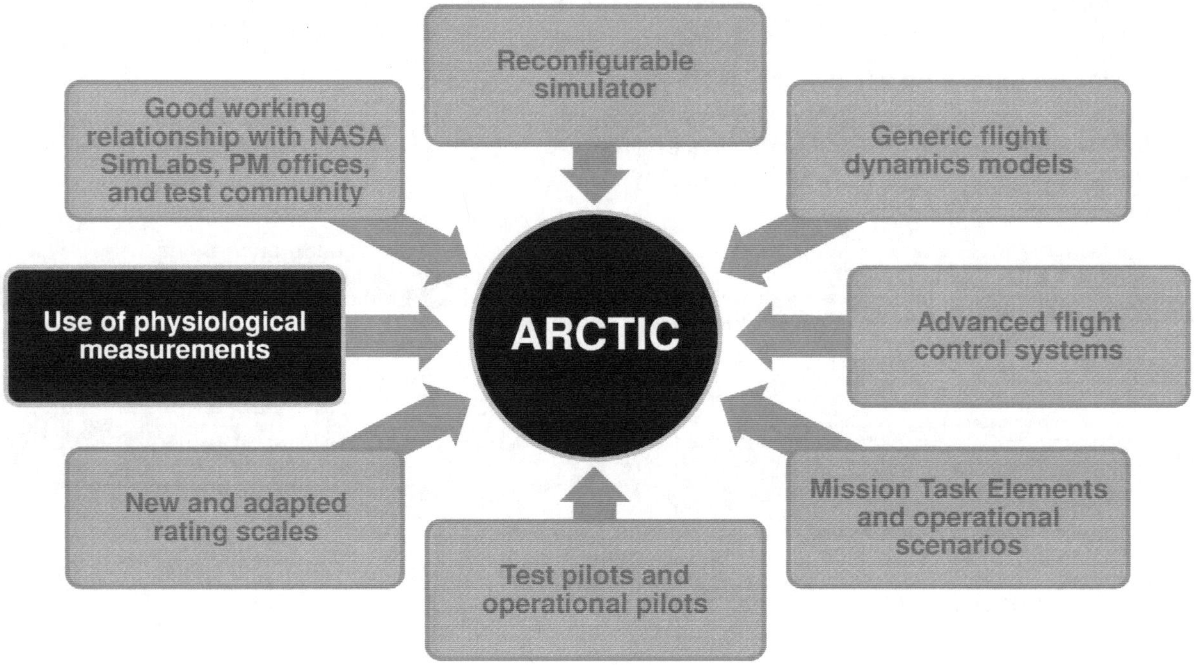

PHYSIOLOGICAL MEASUREMENTS

- **Physiological measurements used to estimate workload and cognitive engagement**
  - Supplement self assess workload ratings
- **Electroencephalogram (EEG) / Electroocoulogram (EOG)**
  - System: B-Alert X10
  - Measures Alpha, Beta, and Theta brain waves
  - BAT – Beta/(Alpha+Theta) ratio: measure of cognitive engagement
- **Heart Rate Monitor**
  - System: Polar H10
  - Measures heart rate variability (HRV)
- **Recorded data time synced to sim data**

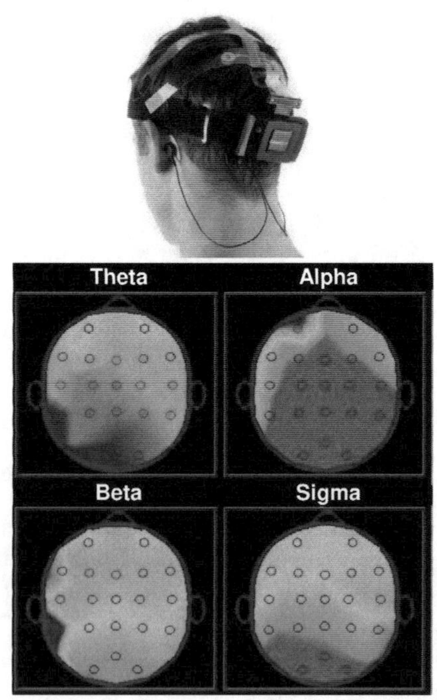

PHYSIOLOGICAL MEASUREMENTS EXAMPLE RESULTS

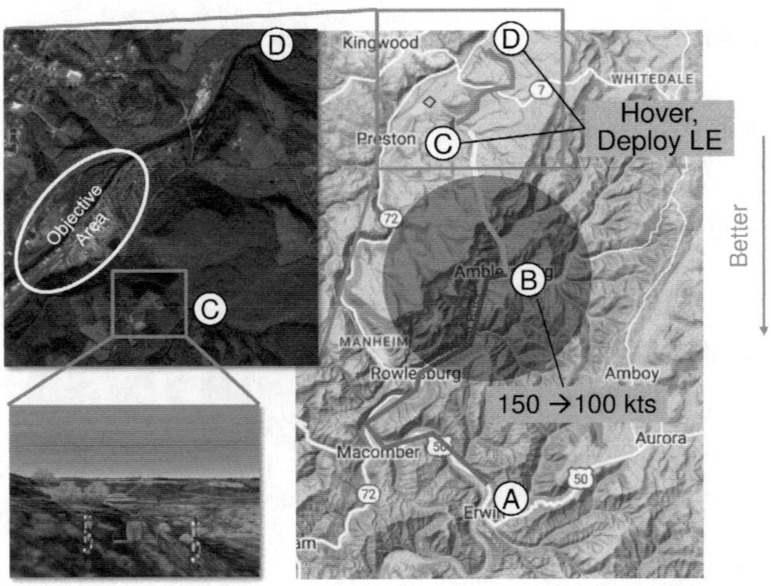

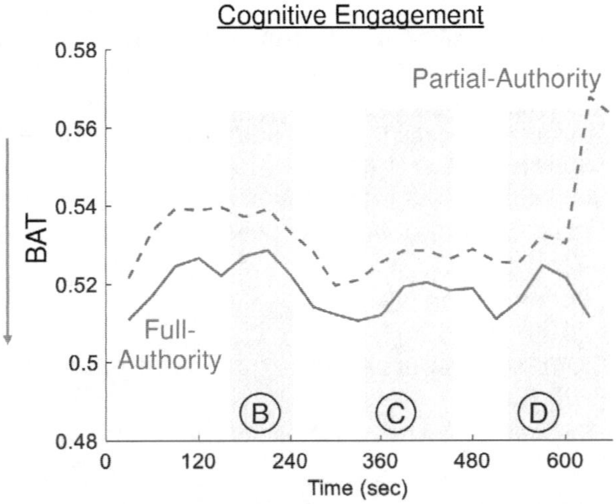

Less workload and cognitive engagement for Full-Authority augmentation measured by EEG

SUMMARY & WHAT'S NEXT

SUMMARY

DEVCOM AVIATION & MISSILE CENTER

Having an agile and independent Government FVL simulation environment has facilitated direct, organic, and indirect experimentation and transition paths:

- **Direct support of PMs to address trade-offs/technical questions with quantitative data, and inform updated requirements**
- **Development of MTEs to assess assigned handling qualities and verify aircraft mission suitability**
- **Collaboration with OEMs**
- **Technical papers summarizing control law design methods and new requirements**
- **Exposure of Army aviators to FVL representative configurations for operationally relevant feedback**

ARCTIC-6: SPRING 2025

DEVCOM AVIATION & MISSILE CENTER

- **Primary Objective:** Introduce autonomous flight behaviors into tiltrotor operating in tactical scenarios in complex terrain

Autonomous Route Planning & Obstacle Avoidance

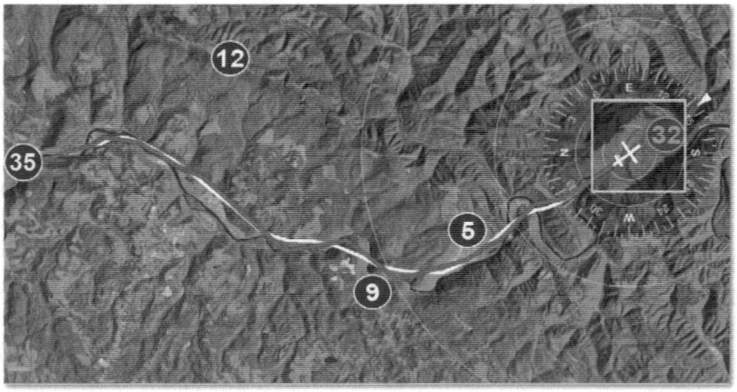

Human-Machine Interface (HMI)

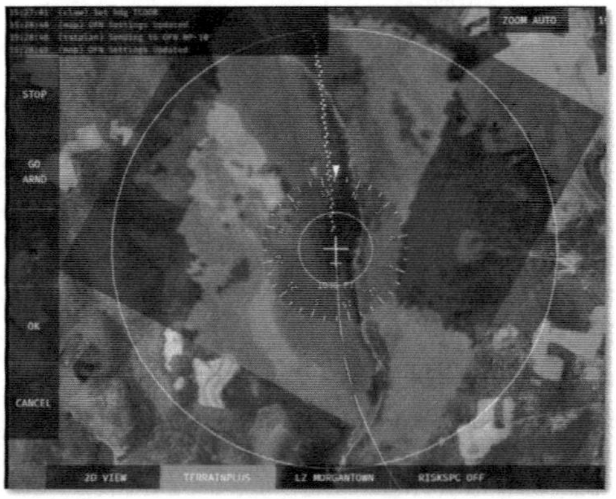

KEY ARCTIC ENABLERS

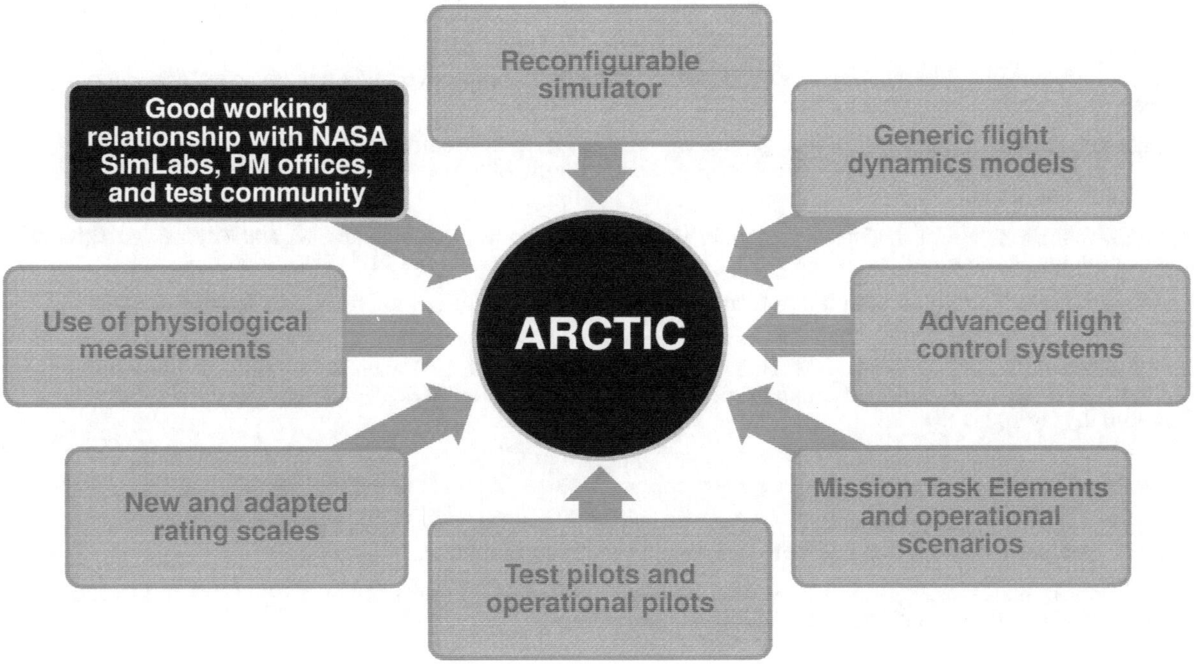

APPROVED FOR PUBLIC RELEASE

APPROVED FOR PUBLIC RELEASE

THANK YOU!

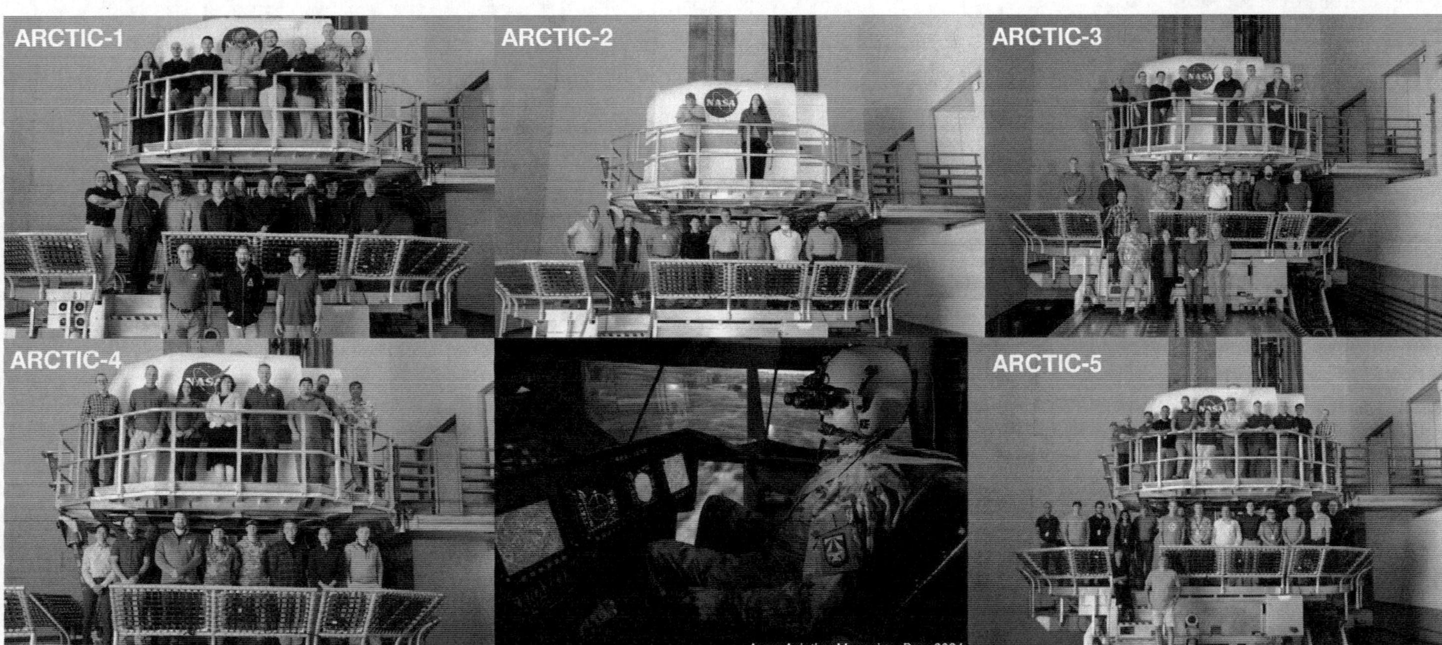

APPROVED FOR PUBLIC RELEASE

ARCTIC PUBLICATIONS

1. Berger, T., Christensen, K., Bail, S., Rominger, J., Clark, S., "**Evaluation of Active Sidestick Failure Mode Handling Qualities**," VFS 2025 HQ Tech Meeting, January 2025
2. D'Alessandro, M., Mackie, R., Berger, T., Ott, C., Sullivan, C. and Curry, I., "**Real-Time Neurophysiological and Subjective Indices of Cognitive Engagement in High-Speed Flight**," *Aerospace Medicine and Human Performance*, 95(12), pp.885-896, 2024
3. Berger, T., Mansur, M., Lusardi, J., Ott, C., Ogden, W., Curry, I., Sullivan, C., D'Alessandro, M. and Mackie, R., "**Effects Of Control System Augmentation Level On Pilot Workload And Effectiveness**," 50th ERF, September 2024
4. Jusko, T. and Berger, T., "**Piloted Simulation Evaluation of MTEs for the Assessment of Low-Level Handling Qualities**," VFS 80th Annual Forum, May 2024
5. Berger, T., Gong, A., Lusardi, J., Mansur, M., Boehringer, R., Ogden, W., Ott, C., Brown, D., Daniels, D., Borden, C., "**Proposed Yaw-Axis Target Acquisition and Tracking Handling Qualities Requirement Updates**," VFS 80th Annual Forum, May 2024
6. Horn, J., Bridges, D., Kuhn, N., Hagwood, D., Berger, T., Gong, A., "**Piloted Simulation Evaluation of Maneuver Optimization Control for a Coaxial Compound Helicopter**," VFS 79th Annual Forum, May 2023
7. Berger, T., Gong, A., Bridges, D., Kuhn, N., Hagwood, D., Horn, J., "**Piloted Simulation Evaluation of Damage Tolerant Control for a Coaxial Compound Helicopter**," VFS 79th Annual Forum, May 2023
8. Berger, T., Dotson, B., Gong, A., Lusardi, J., Mansur, M., Ott, C., Miller, M., "**Effects of Linked vs. Unlinked Cyclic Controllers on Crew Coordination**," VFS 79th Annual Forum, May 2023

THANK YOU.

U.S. ARMY

NRC·CNRC NRC.CANADA.CA

Flight Loads and Aeromechanics Characterization of a Soft Inplane Hingeless Rotor Control System during Handling Qualities Testing

NRC Flight Research Laboratory (Ottawa, Canada)
Publication: VFS.HQ.TM 2025
Marc Alexander, Senior Flight Research Officer
January 2025

National Research Council Canada Conseil national de recherches Canada Canada

Overview

- Rationale, Objectives, Methodology, & Constraints
- Vehicle & Measurement Systems Descriptions
- Flight Test Data Set
- Study Observations:
  - Impacts of MTE's on Loads and Aeromechanics
  - Loads Inter-relationships with Flight Control Augmentation Parameters
  - Loads Variability due to Augmentation and Airspeed
- Summative Discussion on Implications
- Closing Remarks

Research Objectives

Application NRC variable stability helicopter flight research facilities:

- ❑ To study the *interrelationship between flight control system parameters and loads* propagating through the NRC Bell 412 rotor-control system.

Research Rationale

Inter-relationships: Handling Qualities, Control Laws, Life Cycle Parameters

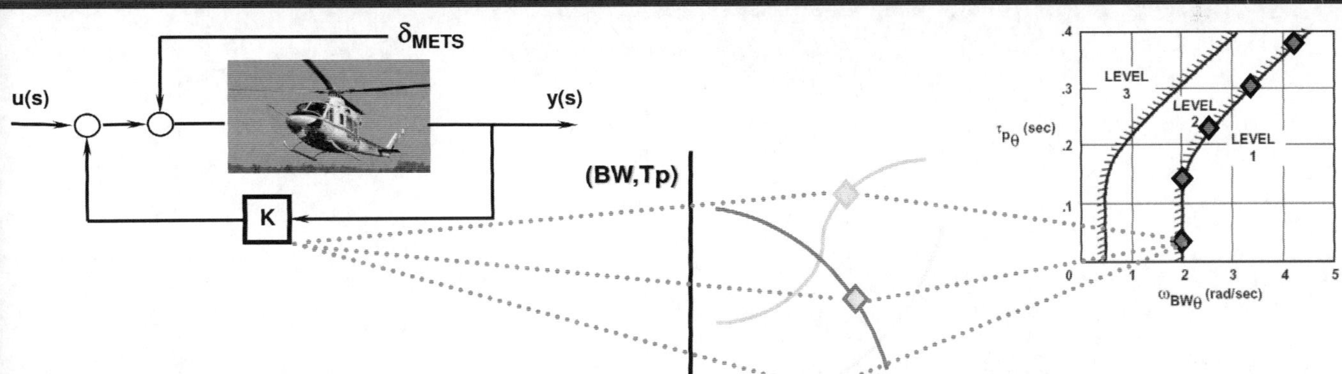

LIFE CYCLE PROBLEM:
- Aircraft Configuration (CG, External Kits)
- Augmentation (a. Response Types, b. Modes)
- Mission Tasks (Agility, Precision, Range)
- Environment (Visual (UCE) & Atm. (Turbulence))
- Loading Conditions (Usage Spectrums)

FLIGHT CONTROL	Rozak
FLYING & HANDLING	Tischler
LIFE CYCLE	Augustin
AIRLOADS	Boussmann
ROTOR DYNAMICS	Kufeld

Research Methodology & Constraints

FLMS: Integration, Validation, Load Surveys

Data Set Selection
- Research: Rotor-Body Dynamics, Lateral Axis
- Excitation Task: Roll-Axis Frequency Sweep
- Mission Tasks: High Speed (Roll Reversal)

Loads to Control Augmentation Analytics
- Response Types: Direct Drive, TRC, RTD, RCAH
- Control Parameters: Bandwidth, Phase Delay, Crossover Frequency)

Research Constraints:
- Limited Data Set (Maneuvers, Flight Hours)
- Limited Data Content (Measurements)
- Single Test Pilot Experimentation
- Unique Test Aircraft
- Experimental Instrumentation

Research Outcome:
- *Test Case Analyses & Observations*
- Not Life Cycle Management Guidance

Systems Descriptions: Test Aircraft and Measurement Systems

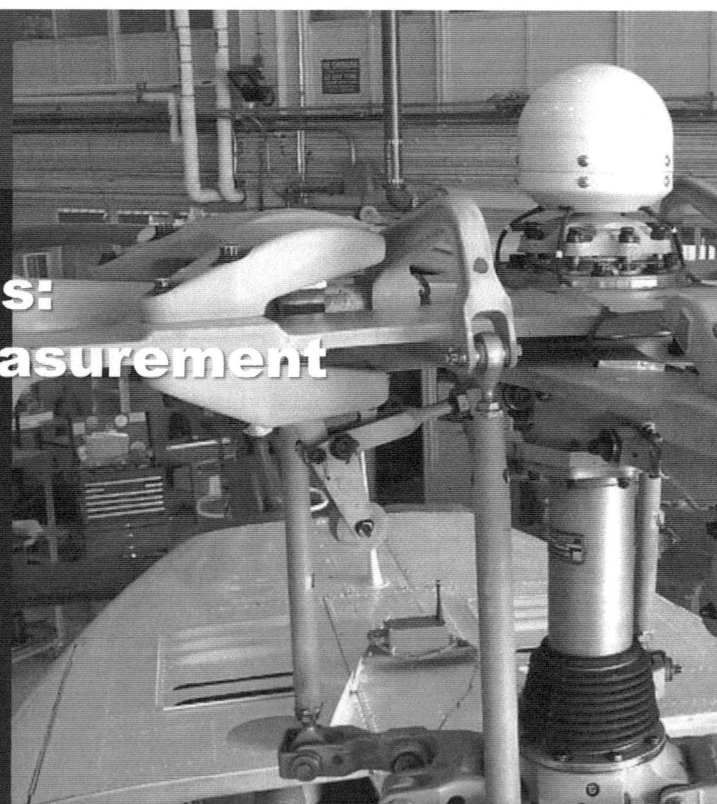

Test Aircraft: NRC Bell 412 Advanced Systems Research Aircraft

Test Aircraft: NRC Bell 412 ASRA

- Capabilities:
  In-Flight Simulation, Variable Stability, Flight Automation, Flight Autonomy
- Flight Control Architecture:
  Simplex with Full Authority Actuation.
- Envelope:
  Full flight envelope engagement.
- Core Safety System:
  Based on Safety pilot.
- Secondary Safety System:
  FBW Computing, Monitoring Software

Inceptors

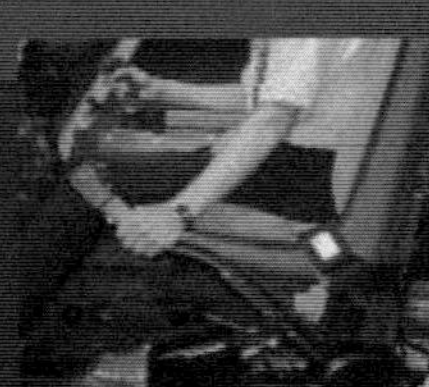

Man-Machine Interface Systems

Safety Pilot Concept

Flight Computing

Flight Software

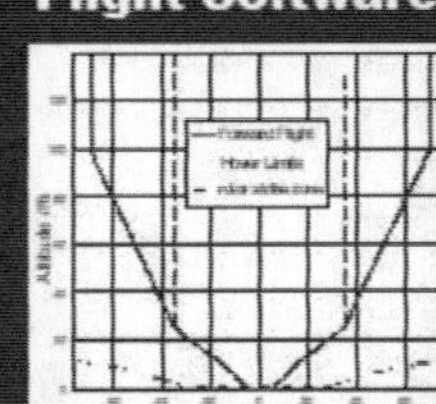

Actuator Technology

State Measurement

NRC Bell412 ASRA Instrumentation: Flight Loads Measurement Suite (FLMS)

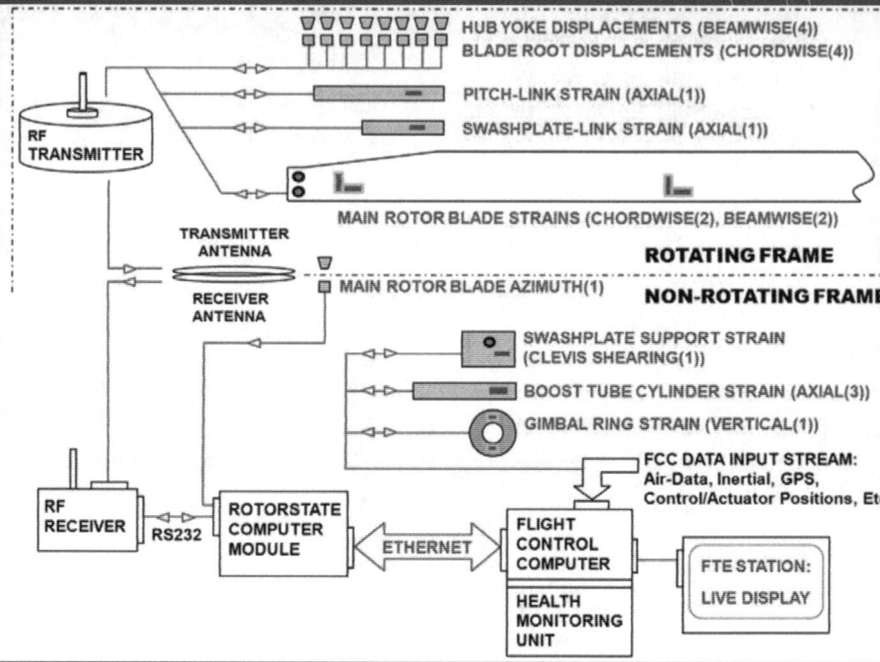

NRC Bell 412: Soft-Inplane Hingeless Rotor Hub

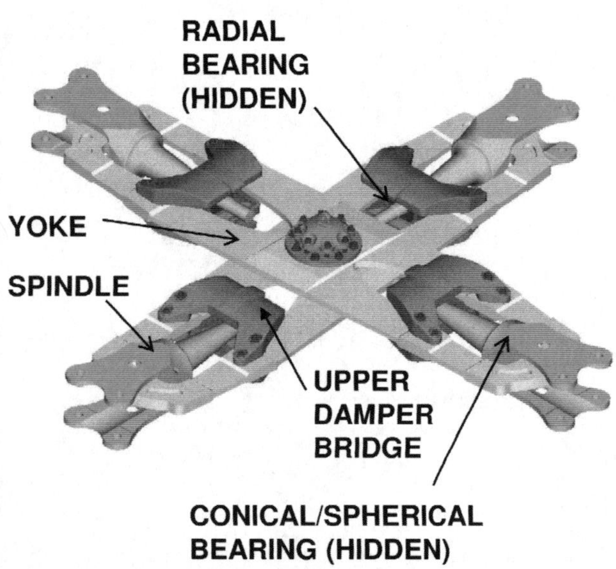

Rotor Diameter	48 ft
Flap Hinge Offset	2.5%R
Lag Hinge Offset	9%R
Pitch Mechanism	Elastomeric
Damper Type	Elastomeric
Rotor Kinematics	Flap-Lag-Pitch

Frequency Placement

	Bell 412HP	BO-105
1st Inplane	0.65/rev	0.73/rev
1st Out-of-Plane	1.03/rev	1.13/rev
2nd Out-of-Plane	2.59/rev	2.78/rev
1st Torsion	5.15/rev	3.13/rev

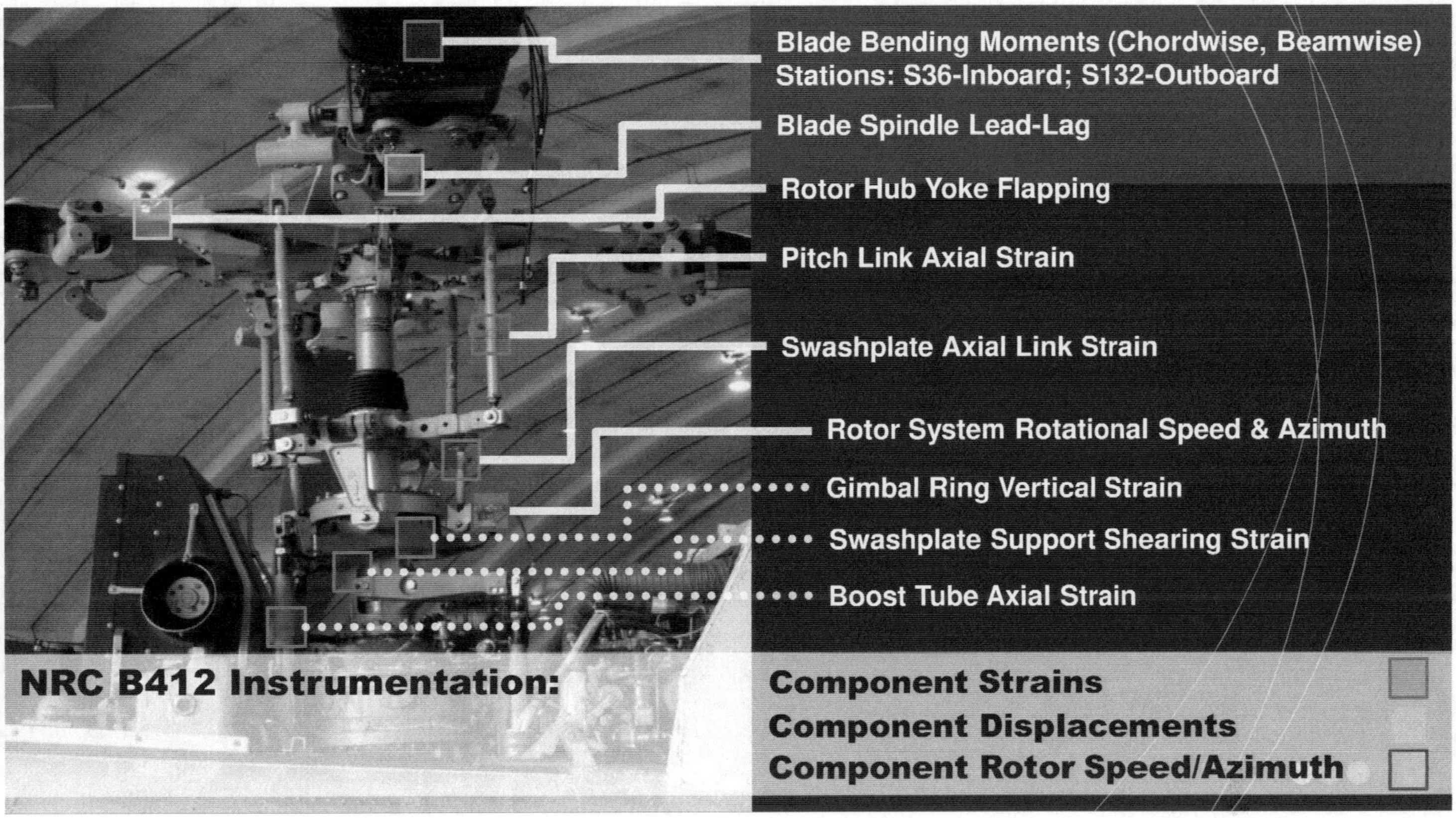
Blade Bending Moments (Chordwise, Beamwise)
Stations: S36-Inboard; S132-Outboard
Blade Spindle Lead-Lag
Rotor Hub Yoke Flapping
Pitch Link Axial Strain
Swashplate Axial Link Strain
Rotor System Rotational Speed & Azimuth
Gimbal Ring Vertical Strain
Swashplate Support Shearing Strain
Boost Tube Axial Strain
NRC B412 Instrumentation:
Component Strains
Component Displacements
Component Rotor Speed/Azimuth

Flight Test Survey:
Test Case and Data Selection

Flight Control Configuration: Response Types

UCE	Control Axes (Roll, Pitch)[c,d]	Notation
GVE[a]	Direct Drive (Open Loop)	DDRV
GVE	Rate Damped	RTD (1,2)
GVE	Rate Command Attitude Hold	RCAH
DVE	Translational Rate Command	TRTC

[a] Good Visual Environment (GVE)
[b] Degraded Visual Environment (DVE)
[c] Yaw Response: Rate Damped (RD)
[d] Heave Response: Open-Loop, Damped, Height-Hold (HTH)

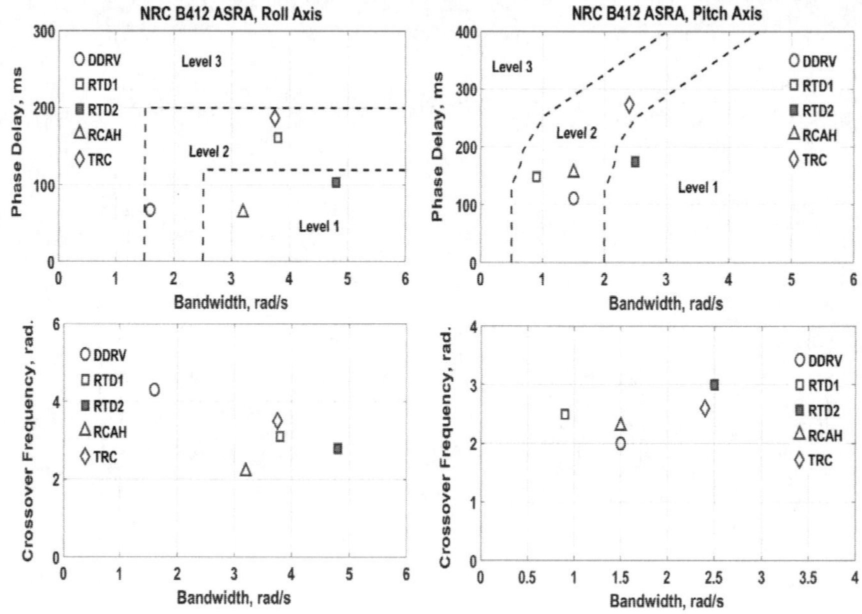

Data: Quality and Set Selection

Parameters	Parameter Ranges	
	Hover Sortie	Flight Sortie
Weight Coefficient	[0.05, 0.06]	[0.055, 0.059]
G-Loading	[2.1, -0.3]	[3.404, -0.561]
Advance Ratio	[0.0013, 0.023]	[0.0013,0.013]
MR Rotor RPM	[318.7, 324.5]	[323.7, 328.1]
Max. Speed (kts)	10	105.5
Altitude (ft)	[0, 80]	[0, 1011]
Winds (deg., kts)	(280, 10)	(300, 8)
Pitch Rate(deg./s)	[8, -8.5]	[27.64, -11.66]
Roll Rate (deg./s)	[17.4, -24.4]	[58.99, -52.58]
Yaw Rate (deg./s)	[7.5, -12]	[51.02, -48.41]

NOTES: Crew (2 Pilots/Conv. Pilot Inceptors, 1 FTE);
FBW System (On/Off); Stability Augmentation (Off); Height Hold (Off)

Data: Roll Reversal, Magnitude Response

Roll Reversal:

- Low altitude (50-100 ft. AGL),
- Moderate speed (40-50 knots),
- High normal Gz (3.4 Gz),
- High roll rate (59 deg./sec.).

- Rotating Frame:
  - Beamwise & chordwise blade bending moments increased 1.5 to 4 times.
  - Blade control linkage (swashplate & pitch links) axial loads increased 1 to 1.5 times.
- Fixed Frame:
  - Swashplate support and gimbal ring loads increased from 0.75 to 1.3 times.

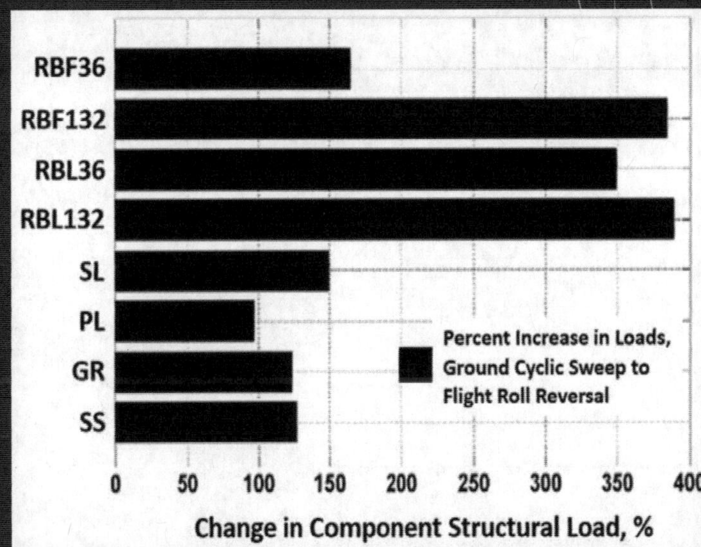

Data: Roll Reversal, Frequency Response

- Rotating Frame:
  - Amplification of 1st chordwise regressive mode (0.65/rev or 3.51 Hz.)
  - Amplification beamwise blade response (3/rev or 16.2 Hz)
- Fixed Frame:
  - Integral (4p/rev) swashplate support loads: induced flapwise moment, lagwise moment, and in-plane shear
  - Non-integral (1/rev, 2/rev, 6/rev) swashplate support: inertial and aerodynamic dissimilarity (instrumented pitch/swashplate links and blades)

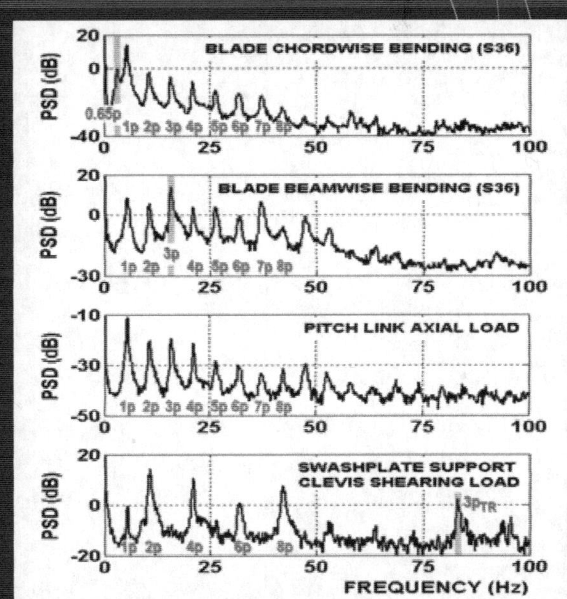

Data Analytics: Assessing Interactions between Loading Conditions and Levels of Control Augmentation

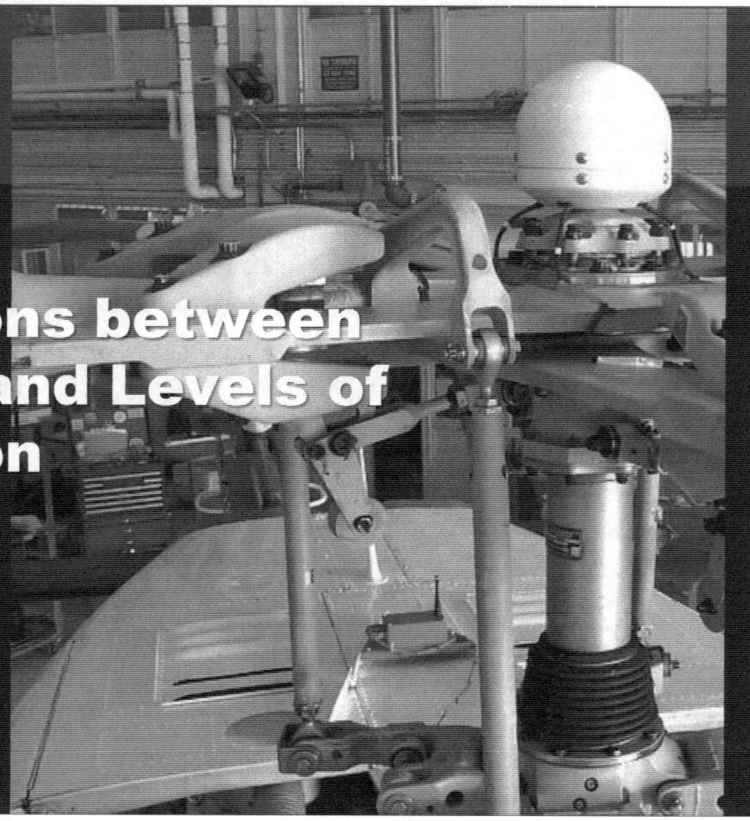

Correlation Study: Bandwidth, Crossover Frequency

ADS-33 Level-1, Roll-axis Boundaries:

- Bandwidth (2.5 to 6 rad/sec.) for phase delays (0 to 125 msec.)

Pilot Crossover Frequency:

- A measure of a pilot's control input activity.

Frequency Sweep Task Bandwidth:

- Estimate = 2 to 3 rad/s.

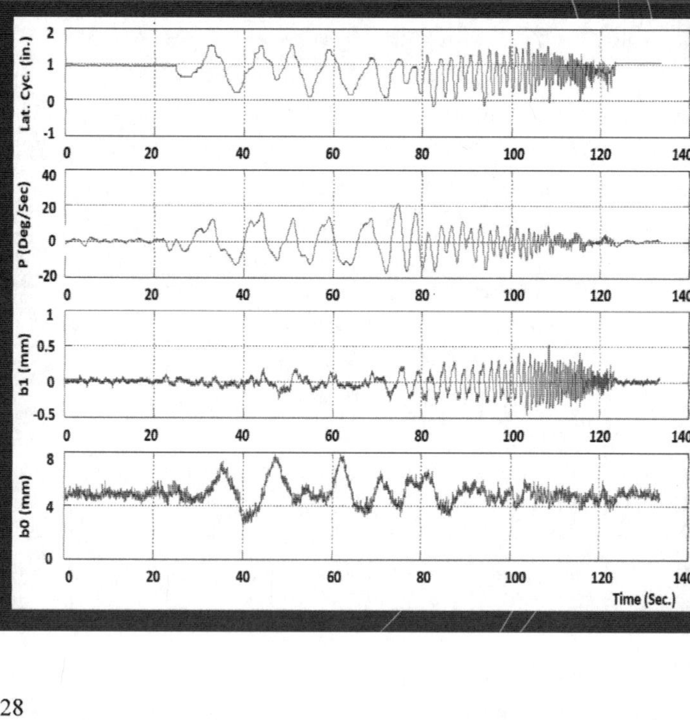

Correlation Study: Bandwidth, Crossover Frequency

ADS-33 Level-1, Roll-axis Boundaries:

- Bandwidth (2.5 to 6 rad/sec.) for phase delays (0 to 125 msec.)

Pilot Crossover Frequency:

- A measure of a pilot's control input activity.

Frequency Sweep Task Bandwidth:

- Estimate = 2 to 3 rad/s.

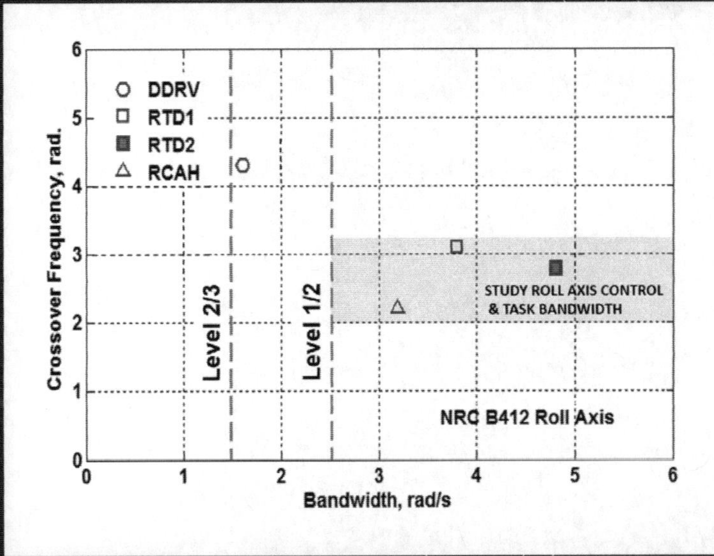

Correlation Study, Hover: Rotor System Loading to Bandwidth

Observations, Load Magnitude:

Fixed Frame:

- SS = Highest Loading;
- GR = Lowest Loading

Rotating Frame:

- RBL.S36 = Highest Loading,
- PL = Lowest Loading

Observations, Response Type:

- RCAH = Highest Loading (Exception: PL)
- DDRV = Lowest Loading

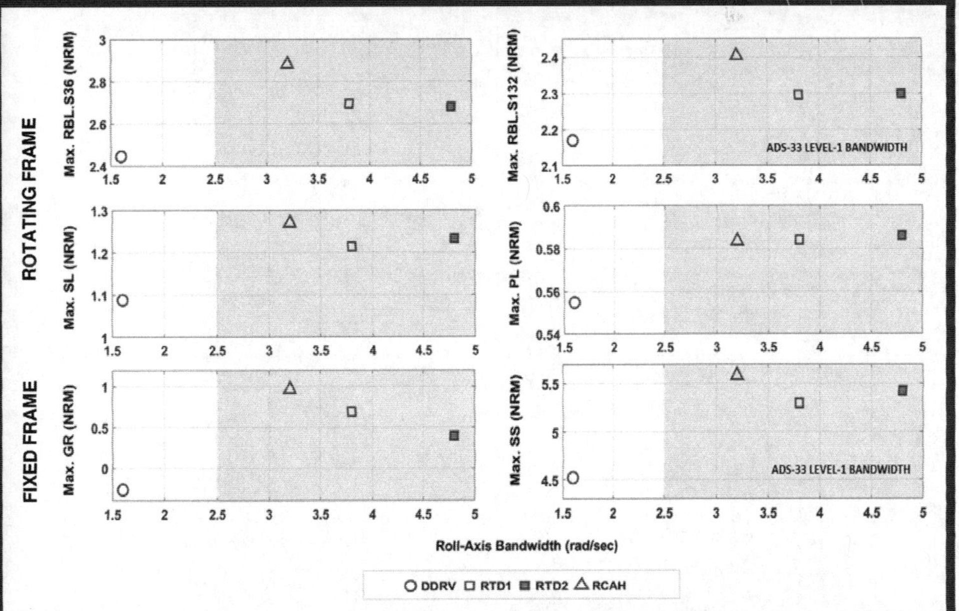

Correlation Study, Forward Flight: Rotor System Loading to Bandwidth

Observations, Load Magnitude; Fixed Frame:

- GR = Highest Loading;
- SS = Lowest Loading

Rotating Frame:

- RBL.S36 = Highest Loading,
- PL = Lowest Loading

Observations, Response Types that may produce the highest loading:

- Rate Damped (RTD1), Direct Drive (DDRV)

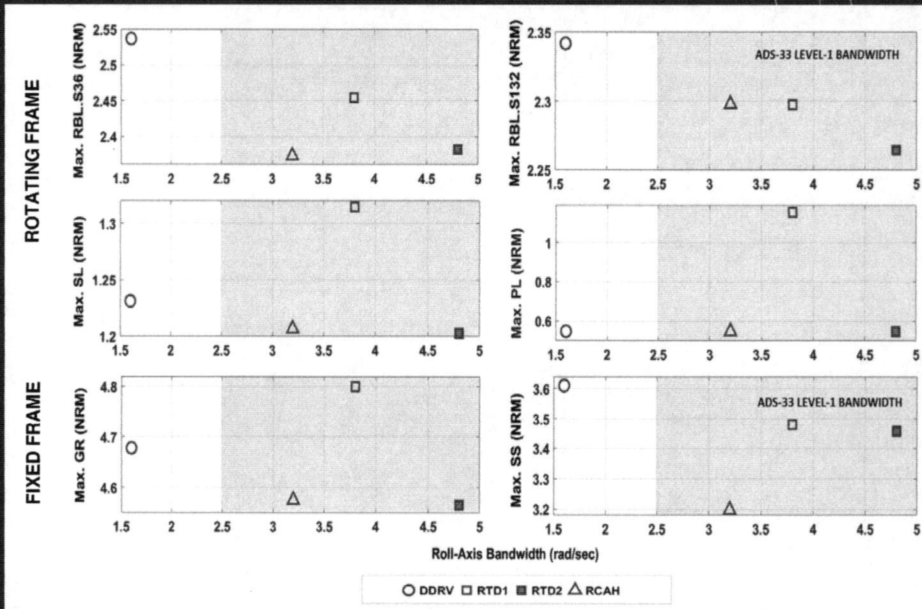

Correlation Study: Rotor System Loading at Crossover, Hover

Observations, Load Magnitude:

Fixed Frame:

- SS = Highest Loading;
- GR = Lowest Loading

Rotating Frame:

- RBL.S36 = Highest Loading,
- PL = Lowest Loading

Observations Relative to Crossover [IC = Inside; OC = Outside] for All (References Frames, Response Types, & Components):

- RCAH = Highest Loading[IC] (Except: PL)
- DDRV = Lowest Loading[OC]

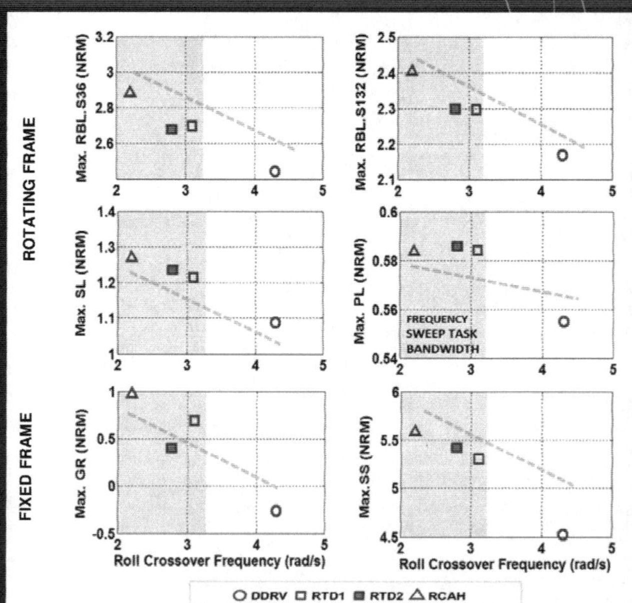

Correlation Study: Rotor System Loading at Crossover, Forward Flight

Observations, Load Magnitude:

Fixed Frame:

- GR = Highest Loading;
- SS = Lowest Loading

Rotating Frame:

- RBL.S36 = Highest Loading,
- PL = Lowest Loading

Observations Relative to Crossover [IC = Inside; OC = Outside] for All (Frames, Response Types, & Components);

Exception: (SS, RBL.S132):

- RTD1 = Highest Loading[IC]
- RCAH or RDT2 = Lowest Loading[IC]

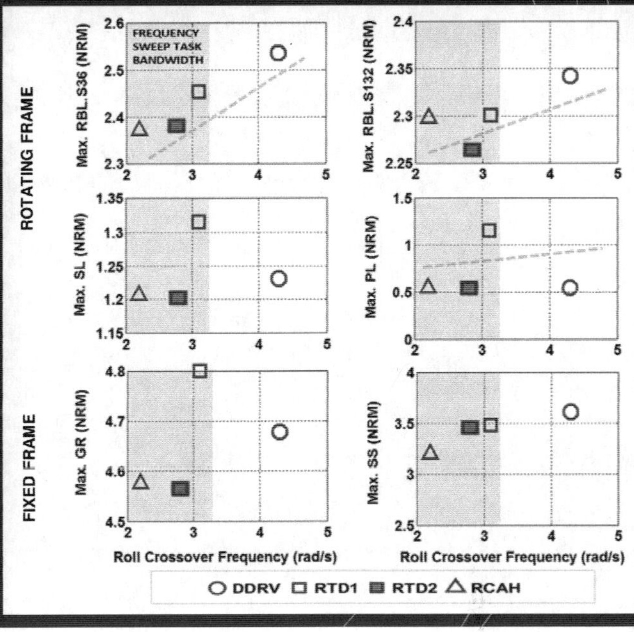

Correlation Study: Flight Speed Variability

Percent load changes between low speed (10kts) and high speed (105kts):

A. Rate-based response types:

- Mixed sign percentage load changes between hover & forward flight.
- Representative of high pilot workload (elevated actuation activity).

B. Attitude-hold response type:

- Greatest load changes between hover & forward flight.
- Representative of aircraft state retention (air-loading as a function of airspeed).

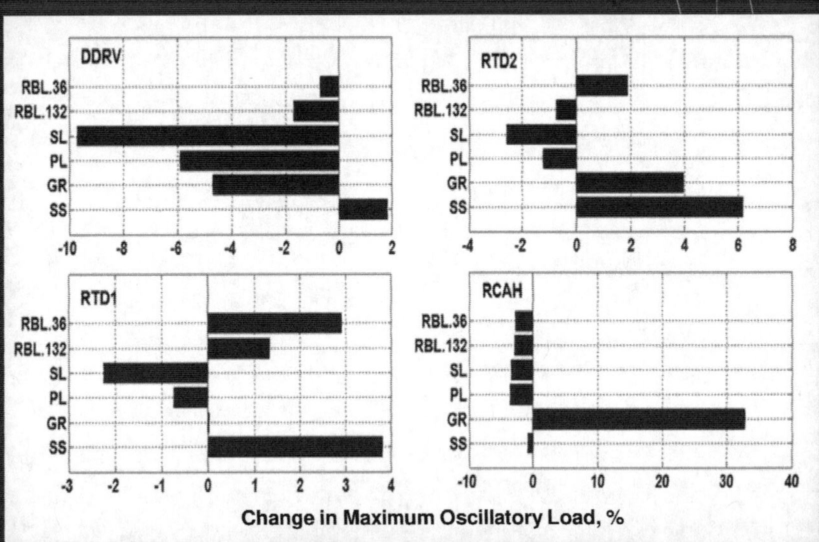

Closing Remarks

Summary and Conclusions

- Mission Task Results: **Amplification of rotor-control system loads with flight dynamics (aerodynamics, Gz, inertia)**
- Correlation Results: **Rate-based responses associated with the highest and lowest loading for hover and forward flight.**
- Airspeed Variability Results: **Rate-based responses types associated with heightened actuation activity (workload). Attitude-hold mode associated with the greatest load changes between hover and forward flight.**

Acknowledgements

'Significant contributions made by ***NRC Flight Research Laboratory*** and the ***Royal Canadian Airforce Aerospace Engineering Test Establishment (RCAF-AETE)*** staff are gratefully acknowledged.'

Questions

Contact:
Marc Alexander,
MarcDavid.Alexander@nrc-cnrc.gc.ca

U.S. ARMY

DEVCOM AVIATION & MISSILE CENTER

U.S. ARMY COMBAT CAPABILITIES DEVELOPMENT COMMAND AVIATION & MISSILE CENTER

Nonlinear Flight Dynamics of Over-Actuated Multi-Lift Rotor eVTOL Aircraft: First Steps

22-23 JAN 2025

Distribution Statement:	Distribution Statement A: Approved for Public Release. Distribution is unlimited.
POC:	Dr. Justin N. Martin

SRD AEROMECHANICS | DR. JUSTIN MARTIN, MODELING AND SIMULATION TEAM
SRD AEROMECHANICS | DR. CHRISTIAN BRACKBILL, MODELING AND SIMULATION TEAM

AGENDA

DEVCOM AVIATION & MISSILE CENTER

- **Background & Motivation**
- RCAS eVTOL Modeling
  - Model development
  - Trim strategies
- Stability & Control
- Future Work

"What is eVTOL?:The Electric Future of Aviation Explained" as presented in PilotMall.com.

https://www.pilotmall.com/blogs/news/what-is-evtol-the-electric-future-of-aviation-explained

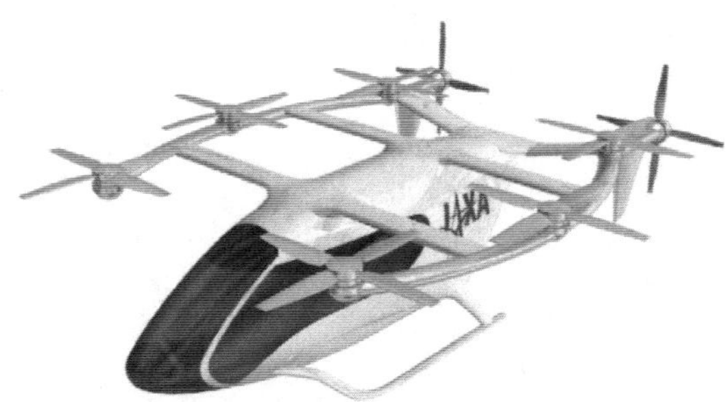

JAXA's Multirotor Lift Offset (MRLO) concept as presented in "Aviation Week"

https://aviationweek.com/aerospace/advanced-air-mobility/jaxas-lift-offset-concept-improves-multicopter-evtol-performance

BACKGROUND + MOTIVATION

- For eVTOL aircraft capable of transporting human pilots/passengers, the air vehicle often includes:
  - –Multiple rotors capable of changing
    - –Rotor speed
    - –Tilt Angle
    - –Blade Pitch
  - –Controllable aerodynamic surfaces such as flaps or ailerons
- New configurations debuted rapidly.
- How do you handle this many DoF?
  - –Over-actuated systems with different control input combinations
  - –Safety
- Repeatability leads to Predictability (best practices)

Image from:
"A Comparison of Three Conceptual Design Approaches Applied to an Electric Distributed Lift Aircraft", 49th ERF Sep 2023.

Approved for Public Release. Distribution Unlimited.

EVTOL MODEL AND ERF AND PLOTS AND STUFF – INTRO2

"Stability and Control Sensitivity of a Lift-Plus-Cruise eVTOL Aircraft to Trim Strategy and Modeling Parameters"

Presented by working group partner at the 50th European Rotorcraft Forum, Marseilles, France 10-12 Sep 2024

U.S./France
Aeromechanics and
Human Factors
Integration Research
(AHFIR)

- Paper outlines:
  - model development
  - RCAS model trim comparison with DynaPy
    - RCAS: U.S. Army code, "Rotorcraft Comprehensive Analysis System"
    - DynaPy: ONERA Flight Dynamics Code
  - Considerations for future studies
- Can now show more Handling Qualities capability.

STABILITY AND CONTROL SENSITIVITY OF A LIFT-PLUS-CRUISE EVTOL AIRCRAFT TO TRIM STRATEGY AND MODELING PARAMETERS

Jean-Paul Reddinger, DEVCOM Army Research Laboratory, Salon de Provence, France
Raphael Perret, ONERA DTIS, Salon de Provence, France
Justin N. Martin, DEVCOM Aviation and Missile Center, Redstone Arsenal, AL, USA
Christian Brackbill, DEVCOM Aviation and Missile Center, Redstone Arsenal, AL, USA
Laurent Binet, ONERA DTIS, Salon de Provence, France

Abstract

This paper presents an evaluation of an eVTOL configuration to provide insight into the stability of the design, depending on the trim strategies used, and accounting for the impact of interactions. The particular eVTOL configuration is an overactuated Lift-Plus-Cruise platform, and the paper focuses initially on examining trim and stability for different sets or groups of control inputs. Control strategies include power-optimal trim, pitch-level trim, low-speed "helicopter mode" and high speed "airplane mode". Primary considerations and tradeoffs between power optimal trim during transition and more controllable trim is attributed to the thrust sensitivity of the RPM controlled rotors. Two different models are used in the trim and stability studies; a low-fidelity flight dynamics analysis with built-in optimization capability, and a medium-fidelity comprehensive analysis. The paper also includes an initial examination of aerodynamic interaction between the wing and the lifting rotors. The impact of wing interference is primarily to increase the down-wash on the rear lifting rotors, which increases the power requirement and reduce the stability of the short period oscillatory mode.

1. INTRODUCTION

Novel electric Vertical Take-Off and Landing (eVTOL) configurations are being developed, with many having reached the prototyping and flight test stages of design. These aircraft benefit from battery and electric motor developments which allow designs that would not have had sufficient range until recently. With such rapid development cycles, many of these configurations have outpaced a rigorous analytical investigation in the open literature. These aircraft introduce stability and control challenges which should be well investigated to ensure safety, reliability, and proper pilot training tools exist prior to their widespread adoption.

The generic Lift-Plus-Cruise (L+C) configuration offers a platform well suited to perform initial studies of several flight modes without the added complexity of moving/tilting propellers. The flight regimes can be loosely grouped into a hover or low-speed "helicopter mode", a high-speed "airplane mode", and a blended control mode for transition and other low-to-moderate-speed maneuvers. Handling qualities analysis and control system design for this configuration is an active research area, especially recently.

Ref. [1, 2, 3] examine piloting issues and handling qualities of over-actuated vehicles. Research focused on L+C control system design and testing includes Ref. [4, 5]. The transition mode control problem is the focus of Ref. [6], while Ref. [7] and Ref. [8] look at more detailed aerodynamic analyses of components such as propellers or wings. Finally, Ref. [9, 8] focus on the overall design of L+C aircraft and various trim strategies. Many of these related efforts are in the early research stages, and in general there is a lack of public domain validation data.

This paper seeks to examine the stability of a representative eVTOL aircraft, using a generic Lift + Cruise configuration introduced in Ref. [9]. Building upon this and similar work Ref. [8, 10], the focus is primarily on the locus of the eigenvalues of the open-loop system/bare airframe, and its evolution with different modeling parameters. The final aim of this stability study is to identify key points of interest concerning the stability of eVTOL configurations

Distribution Statement A: Approved for public release. Distribution is unlimited. Page 1 of 14
Presented at 50th European Rotorcraft Forum, Marseille, France, 10-12 September, 2024.

AGENDA

- Background & Motivation
- **RCAS eVTOL Modeling**
  - **–Model development**
  - **–Trim strategies**
- Stability & Control
- Future Work

Image from:
https://spinoff.nasa.gov/Spinoff2020/t_2.html

Wisk Aero Cora, aviationweek.com

```
S TRIMVAR
! Trim    TrimVar     TrimVar    Target   Target      Target
! VarID  PertValue    DampFac      ID      Value        Tol
!@ this will be for zero pitch attitude w/ flaps trim
a   5       5          0.4         1        0          100.0      ! Throttle, Fx
a   6     0.02         0.4         2        0          100.0      ! Roll, Fy
a  151      5          0.4         3        0          100.0      ! Right mwing flap, Fz
a  152      5          0.4         4        0          100.0      ! Left mwing aileron, Mx
a  153      5          0.4         5        0          100.0      ! htail elevon, My
a  154      5          0.4         6        0          100.0      ! Left Rudder, Mz
!@
!@ this will be for pitch controlled, no flaps trim
!@a   5       5          0.4         1        0          100.0      ! Throttle, Fx
!@a   6      .02         0.4         2        0          100.0      ! Roll, Fy
!@a   7      .02         0.4         3        0          100.0      ! Pitch, Fz
!@a  151      5          0.4         4        0          100.0      ! Left mwing, Mx
!@a  153      5          0.4         5        0          100.0      ! htail elevon, My
!@a  154      5          0.4         6        0          100.0      ! Left Rudder, Mz
```


EVTOL MODEL DEVELOPMENT

US-France AHFIR

- Collaborative Agreement to work rotorcraft issues
- Composed of multiple tasks simultaneously
- Tasks may interconnect
- Some collaboration with non-government participants

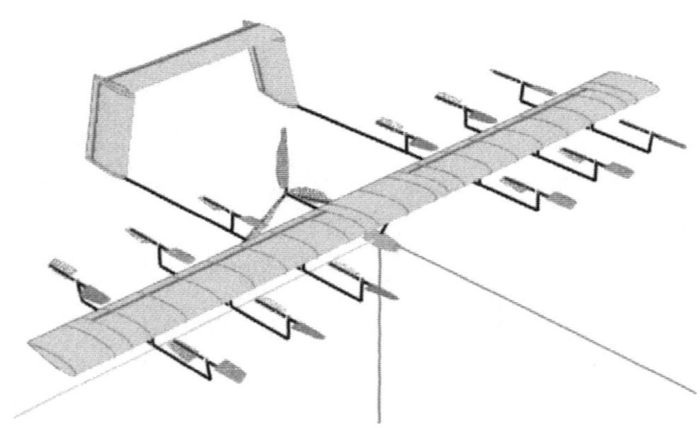

EVTOL MODEL DEVELOPMENT

DEVCOM AVIATION & MISSILE CENTER

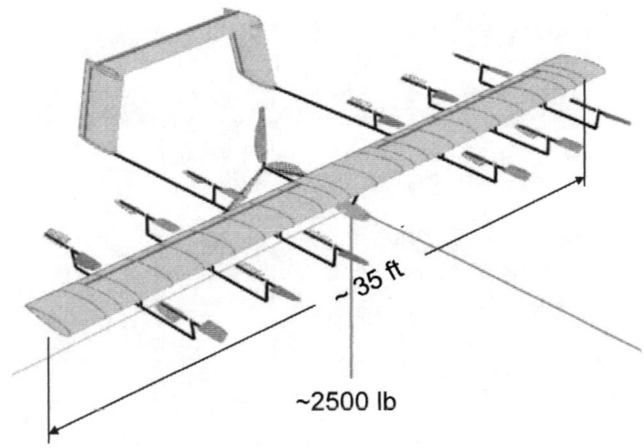

EVTOL MODEL DEVELOPMENT

DEVCOM AVIATION & MISSILE CENTER

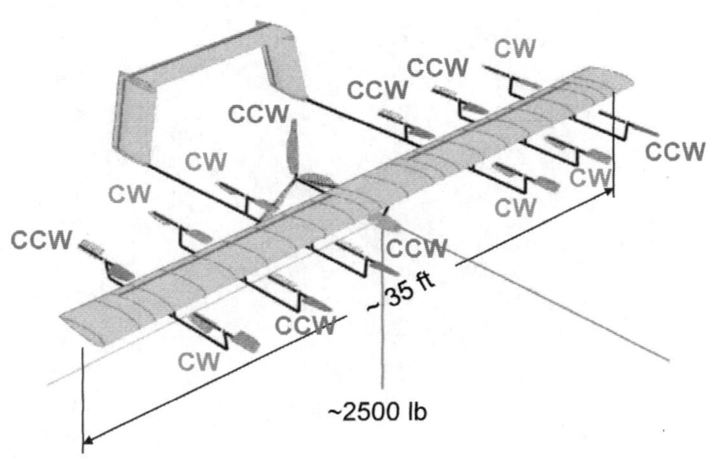

EVTOL MODEL DEVELOPMENT

DEVCOM AVIATION & MISSILE CENTER

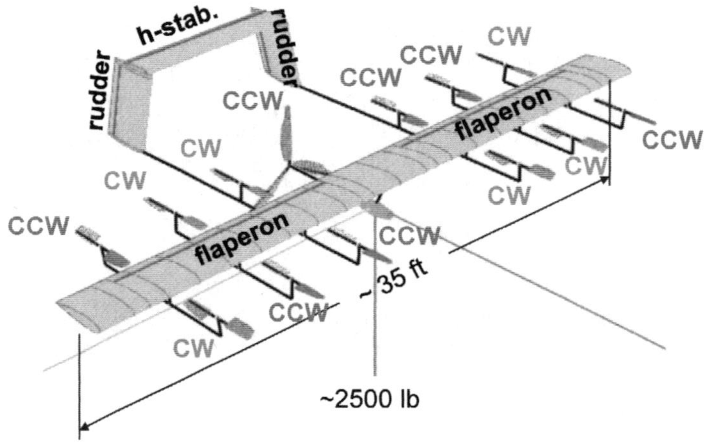

Approved for Public Release. Distribution Unlimited.

EVTOL MODEL DEVELOPMENT

DEVCOM AVIATION & MISSILE CENTER

Higher speed helicopter and airplane mode maneuverability passed to surfaces and propeller.

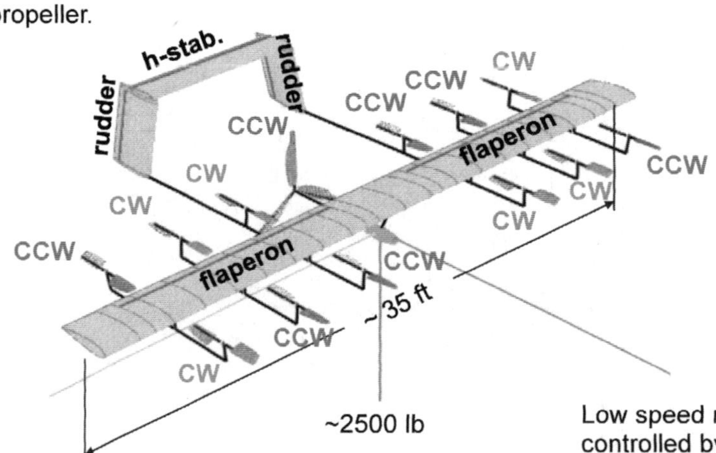

Low speed maneuverability controlled by variable speeds in lifting rotors.

Approved for Public Release. Distribution Unlimited.

TWO CATEGORIES OF TRIM

Unique trim strategies

- **Multirotor** – Grouped differential RPM controls, pitch attitude used
  - All control surfaces set to zero, propeller RPM to zero
  - **Collective** RPM, **Lateral** RPM, **Longitudinal** RPM, and **Pedal** RPM
- Airplane – Airplane mode controls, including propeller and pitch attitude
  - All lift-rotor RPMs set to zero
  - **Differential Flaperons**, **Elevator**, **Rudder**, and **Propeller** Thrust

Over-actuated strategies (optimization required, minimizing power)

- **Pitch-level Optimal** – All controls used
  - pitch attitude fixed at zero degrees

Target	Multirotor Mode	Pitch-Level	Airplane + Flaps (Pitch-Level)
F_x	Pitch Attitude	Propeller RPM	Propeller RPM
F_y	Roll Attitude	Roll Attitude	Roll Attitude
F_z	Collective RPM	Collective RPM	Flaps
M_x	Lateral RPM	Lateral RPM	Ailerons
M_y	Longitudinal RPM	Longitudinal RPM	Elevator
M_z	Pedal RPM	Pedal RPM	Rudder
handling of redundant controls:			
	Propeller Off	Flaps and Elevator Scheduled	Lifting Rotors Off
	Control Surfaces Zeroed	Pitch Attitude Fixed	Pitch Attitude Fixed

Pulled from ERF Paper referenced earlier

Approved for Public Release. Distribution Unlimited.

Approved for Public Release. Distribution Unlimited.

RCAS MIXER AND MIXING

XC XA XB XP

```
S CONTROLMIXER
! Control  Value at Zero    -------- Coefficients for Pilot Control ------
!   ID     Pilot Control    Coll.    Lat.    Long.    Pedal    Throt.
a   1           0             1       +1      -1       +1       0 !@fwd long: aft rotor is +
a   2           0             1       +1      +1       -1       0 !@right lat: left side +
a   3           0             1       +1      -1       -1       0 !@up coll: all +
a   4           0             1       +1      +1       +1       0 !@right ped: ccw +
a   5           0             1       +1      -1       -1       0
a   6           0             1       +1      +1       +1       0
a   7           0             1       -1      -1       +1       0
a   8           0             1       -1      +1       -1       0
a   9           0             1       -1      -1       +1       0
a  10           0             1       -1      +1       -1       0
a  11           0             1       -1      -1       -1       0
a  12           0             1       -1      +1       +1       0
a  13           0             0        0       0        0       1

S CONTROLMIXING
! Row   Aircraft    Conversion or     Pilot/UDF
!  ID   Cont. ID    Mixing Gain       Cont. ID
a  14     14          0.017453          101   ! Lifting Rotors Collective Pitch
a  15     15          0.017453          102   ! Pusher Prop Collective Pitch
```

```
S CONTROLMIXING
! Row   Aircraft    Conversion or     Pilot/UDF
!  ID   Cont. ID    Mixing Gain       Cont. ID
a  14     14          0.017453          101   ! Lifting Rotors Collective Pitch
a  15     15          0.017453          102   ! Pusher Prop Collective Pitch
a  51     51          0.017453          151   !@ Left Wing Flaperon
a  52     52          0.017453          152   !@ Right Wing Aileron
!@c 52      52         -0.017453           151   !@ Right Wing Aileron - Used for flap schedule
a  53     53          0.017453          153   !@ elevator down deflection
a  54     54          0.017453          154   !@ Left Rudder
a  55     55          0.017453          155   !@ Right Rudder
c  55     55          0.017453          154   !@ Right Rudder
```

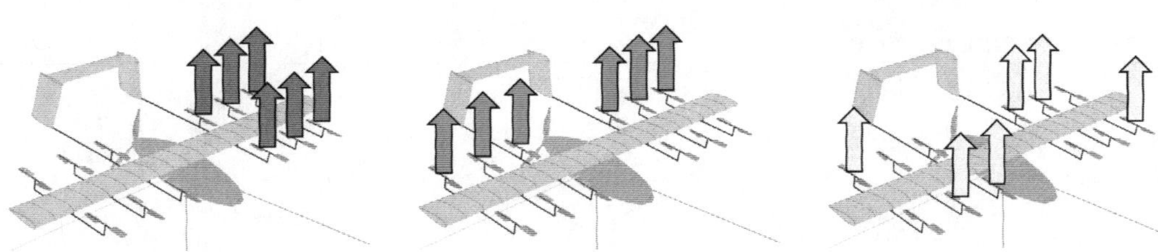

Approved for Public Release. Distribution Unlimited.

FLAPERONS, A NOTE

DEVCOM AVIATION & MISSILE CENTER

- What is a flaperon? Well, for this work…
  - Single control surface on an individual wing.
  - Assists in vertical lift in forward flight with "flap-like" application.
  - Assists in rolling moments in forward flight with "aileron-like" application.
- How did we implement them?
  - RCAS uses a left and right "flap" input for trim
  - (left + right) / 2 = "flaps"
  - (left – right) / 2 = "aileron"
  - Both are controlled independently for trim

Target	Airplane + Flaps (Pitch-Level)
F_x	Propeller RPM
F_y	Roll Attitude
F_z	Flaps
M_x	Ailerons
M_y	Elevator
M_z	Rudder
	Lifting Rotors Off
	Pitch Attitude Fixed

Pulled from ERF Paper referenced earlier

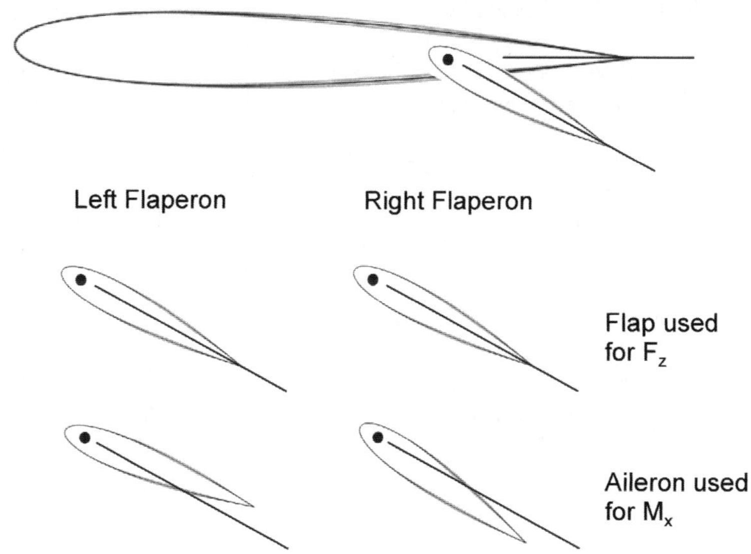

LIFTING ROTOR RPM DIFFERENTIALS

DEVCOM AVIATION & MISSILE CENTER

- Differential Longitudinal RPM: Change in average RPM of lifting rotors to increase speed.
- Uniform vs. 3-state Inflow
- Control Surface use during transition
- Airplane mode begins and ends with lifting rotors seized.

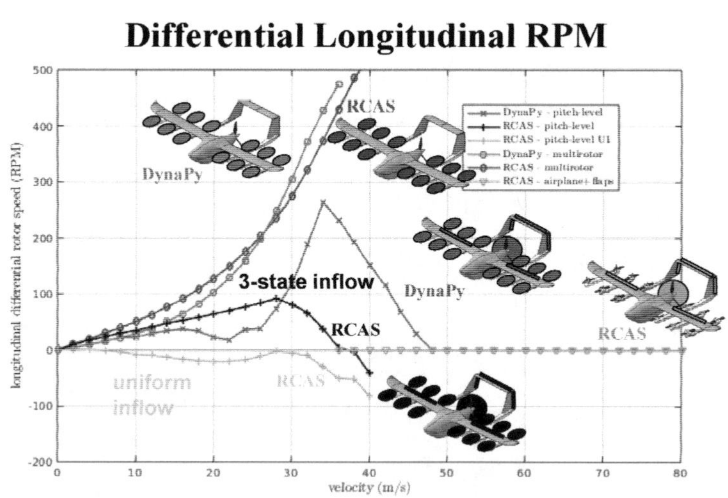

AGENDA

- Background & Motivation
- RCAS eVTOL Modeling
  - Model development
  - Trim strategies
- **Stability & Control**
- Future Work

NASA Crash Tests eVTOL Concepts

https://www.helicoptersmagazine.com/nasa-crash-tests-evtol-concept/

Approved for Public Release. Distribution Unlimited.

Approved for Public Release. Distribution Unlimited.

ROOT LOCUS

- Evolution of modes from hover to 80 m/s
  - All modes in hover are identical
  - Low frequency modes insensitive to trim strategy
- Roll subsidence mode
  - More stable rolling motion with increased velocity
- Short period mode
  - Natural frequency / stability increase with speed
- Dutch roll mode
  - Multi-rotor mode was found to be less stable (not shown)
- Little difference between RCAS's Uniform Inflow and 3-state Inflow models
  - Largest difference is in the roll subsidence stability around hover
- What's up with the gaps between flight modes?

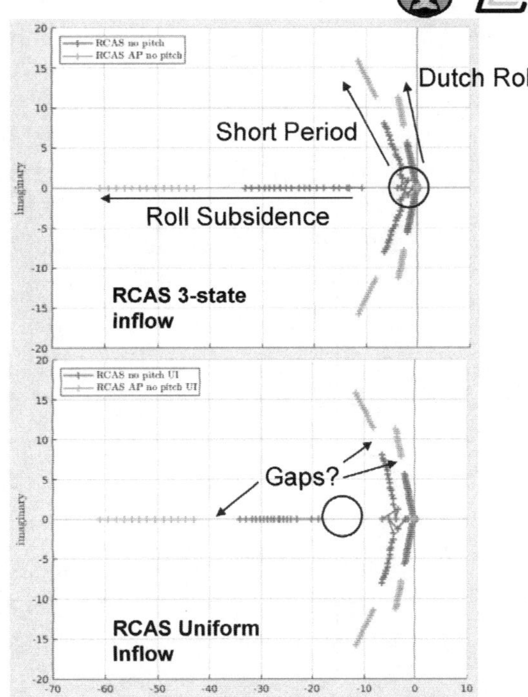

Approved for Public Release. Distribution Unlimited.

WHY IS THE RCAS CONTROL DERIVATIVE A LITTLE DIFFERENT WITH EVTOL?

Hover	Low Speed	Transition	High Speed
Longitudinal Stick	Longitudinal Stick	Longitudinal Stick	Throttle
Lateral Stick	Lateral Stick	Lateral Stick	Flaps
Collective	Collective	Collective	Ailerons
Pedal	Pedal	Pedal	Rudder
Pitch Attitude	Pitch Attitude	Pitch Attitude	Horizontal Tail
Roll Attitude	Roll Attitude	Roll Attitude	
	Throttle	Throttle	
		Flaps	
		Ailerons	
		Rudder	
		Horizontal Tail	

- 17 controls in our case
- A control system will need to mix these regimes based on flight conditions
- RCAS Stability and Control functions track these, whether used or not.

CONTROL DERIVATIVE MATRIX FROM RCAS

	XC	XA	XB	XP	THROT	Lift Rotor Col. Pitch	Propeller Col. Pitch	FLAP	AILERON	H-TAIL FLAP	RUDDER (L)	RUDDER (R)
F_x	$X_{\delta,col}$	$X_{\delta,lat}$	$X_{\delta,lng}$	$X_{\delta,ped}$	$X_{\delta,thrt}$	$X_{\delta,LRc}$	$X_{\delta,Pc}$	$X_{\delta,flap}$	$X_{\delta,ail}$	$X_{\delta,HT}$	$X_{\delta,rud}$	$X_{\delta,rud}$
F_y	$Y_{\delta,col}$	$Y_{\delta,lat}$	$Y_{\delta,lng}$	$Y_{\delta,ped}$	$Y_{\delta,thrt}$	$Y_{\delta,LRc}$	$Y_{\delta,Pc}$	$Y_{\delta,flap}$	$Y_{\delta,ail}$	$Y_{\delta,HT}$	$Y_{\delta,rud}$	$Y_{\delta,rud}$
F_z	$Z_{\delta,col}$	$Z_{\delta,lat}$	$Z_{\delta,lng}$	$Z_{\delta,ped}$	$Z_{\delta,thrt}$	$Z_{\delta,LRc}$	$Z_{\delta,Pc}$	$Z_{\delta,flap}$	$Z_{\delta,ail}$	$Z_{\delta,HT}$	$Z_{\delta,rud}$	$Z_{\delta,rud}$
M_x	$L_{\delta,col}$	$L_{\delta,lat}$	$L_{\delta,lng}$	$L_{\delta,ped}$	$L_{\delta,thrt}$	$L_{\delta,LRc}$	$L_{\delta,Pc}$	$L_{\delta,flap}$	$L_{\delta,ail}$	$L_{\delta,HT}$	$L_{\delta,rud}$	$L_{\delta,rud}$
M_y	$M_{\delta,col}$	$M_{\delta,lat}$	$M_{\delta,lng}$	$M_{\delta,ped}$	$M_{\delta,thrt}$	$M_{\delta,LRc}$	$M_{\delta,Pc}$	$M_{\delta,flap}$	$M_{\delta,ail}$	$M_{\delta,HT}$	$M_{\delta,rud}$	$M_{\delta,rud}$
M_z	$N_{\delta,col}$	$N_{\delta,lat}$	$N_{\delta,lng}$	$N_{\delta,ped}$	$N_{\delta,thrt}$	$N_{\delta,LRc}$	$N_{\delta,Pc}$	$N_{\delta,flap}$	$N_{\delta,ail}$	$N_{\delta,HT}$	$N_{\delta,rud}$	$N_{\delta,rud}$

CONTROL DERIVATIVE MATRIX FROM RCAS

	XC	XA	XB	XP	THROT	Lift Rotor Col. Pitch - Fixed	Propeller Col. Pitch - Fixed	FLAP	AILERON	H-TAIL FLAP	RUDDER (L)	RUDDER (R) L-RUD CNTL
F_x	$X_{\delta,col}$	$X_{\delta,lat}$	$X_{\delta,lng}$	$X_{\delta,ped}$	$X_{\delta,thrt}$	~~$X_{\delta,LRc}$~~	~~$X_{\delta,PRc}$~~	$X_{\delta,flap}$	$X_{\delta,ail}$	$X_{\delta,HT}$	↓ $X_{\delta,rud}$	~~$X_{\delta,rud}$~~
F_y	$Y_{\delta,col}$	$Y_{\delta,lat}$	$Y_{\delta,lng}$	$Y_{\delta,ped}$	$Y_{\delta,thrt}$	~~$Y_{\delta,LRc}$~~	~~$Y_{\delta,PRc}$~~	$Y_{\delta,flap}$	$Y_{\delta,ail}$	↓ $Y_{\delta,HT}$	$Y_{\delta,rud}$	~~$Y_{\delta,rud}$~~
F_z	$Z_{\delta,col}$	$Z_{\delta,lat}$	$Z_{\delta,lng}$	$Z_{\delta,ped}$	$Z_{\delta,thrt}$	~~$Z_{\delta,LRc}$~~	~~$Z_{\delta,PRc}$~~	$Z_{\delta,flap}$	$Z_{\delta,ail}$	$Z_{\delta,HT}$	↓ $Z_{\delta,rud}$	~~$Z_{\delta,rud}$~~
M_x	$L_{\delta,col}$	$L_{\delta,lat}$	$L_{\delta,lng}$	$L_{\delta,ped}$	$L_{\delta,thrt}$	~~$L_{\delta,LRc}$~~	~~$L_{\delta,PRc}$~~	$L_{\delta,flap}$	$L_{\delta,ail}$	↓ $L_{\delta,HT}$	$L_{\delta,rud}$	~~$L_{\delta,rud}$~~
M_y	$M_{\delta,col}$	$M_{\delta,lat}$	$M_{\delta,lng}$	$M_{\delta,ped}$	$M_{\delta,thrt}$	~~$M_{\delta,LRc}$~~	~~$M_{\delta,PRc}$~~	$M_{\delta,flap}$	$M_{\delta,ail}$	$M_{\delta,HT}$	↓ $M_{\delta,rud}$	~~$M_{\delta,rud}$~~
M_z	$N_{\delta,col}$	$N_{\delta,lat}$	$N_{\delta,lng}$	$N_{\delta,ped}$	$N_{\delta,thrt}$	~~$N_{\delta,LRc}$~~	~~$N_{\delta,PRc}$~~	$N_{\delta,flap}$	$N_{\delta,ail}$	↓ $N_{\delta,HT}$	$N_{\delta,rud}$	~~$N_{\delta,rud}$~~

CONTROL DERIVATIVE MATRIX FROM RCAS

	XC	XA	XB	XP	THROT	Lift Rotor Col. Pitch - Fixed	Propeller Col. Pitch - Fixed	FLAP	AILERON	H-TAIL FLAP	RUDDER (L)	RUDDER (R) L-RUD CNTL
F_x	$X_{\delta,col}$	$X_{\delta,lat}$	$X_{\delta,lng}$	$X_{\delta,ped}$	**$X_{\delta,thrt}$**	~~$X_{\delta,LRc}$~~	~~$X_{\delta,PRc}$~~	$X_{\delta,flap}$	$X_{\delta,ail}$	$X_{\delta,HT}$	↓ $X_{\delta,rud}$	~~$X_{\delta,rud}$~~
F_y	$Y_{\delta,col}$	$Y_{\delta,lat}$	$Y_{\delta,lng}$	$Y_{\delta,ped}$	$Y_{\delta,thrt}$	~~$Y_{\delta,LRc}$~~	~~$Y_{\delta,PRc}$~~	$Y_{\delta,flap}$	$Y_{\delta,ail}$	↓ $Y_{\delta,HT}$	$Y_{\delta,rud}$	~~$Y_{\delta,rud}$~~
F_z	**$Z_{\delta,col}$**	$Z_{\delta,lat}$	$Z_{\delta,lng}$	$Z_{\delta,ped}$	$Z_{\delta,thrt}$	~~$Z_{\delta,LRc}$~~	~~$Z_{\delta,PRc}$~~	$Z_{\delta,flap}$	$Z_{\delta,ail}$	$Z_{\delta,HT}$	↓ $Z_{\delta,rud}$	~~$Z_{\delta,rud}$~~
M_x	$L_{\delta,col}$	**$L_{\delta,lat}$**	$L_{\delta,lng}$	$L_{\delta,ped}$	$L_{\delta,thrt}$	~~$L_{\delta,LRc}$~~	~~$L_{\delta,PRc}$~~	**$L_{\delta,flap}$**	**$L_{\delta,ail}$**	↓ $L_{\delta,HT}$	$L_{\delta,rud}$	~~$L_{\delta,rud}$~~
M_y	$M_{\delta,col}$	$M_{\delta,lat}$	**$M_{\delta,lng}$**	$M_{\delta,ped}$	$M_{\delta,thrt}$	~~$M_{\delta,LRc}$~~	~~$M_{\delta,PRc}$~~	**$M_{\delta,flap}$**	$M_{\delta,ail}$	**$M_{\delta,HT}$**	↓ $M_{\delta,rud}$	~~$M_{\delta,rud}$~~
M_z	$N_{\delta,col}$	$N_{\delta,lat}$	$N_{\delta,lng}$	**$N_{\delta,ped}$**	$N_{\delta,thrt}$	~~$N_{\delta,LRc}$~~	~~$N_{\delta,PRc}$~~	$N_{\delta,flap}$	$N_{\delta,ail}$	↓ $N_{\delta,HT}$	**$N_{\delta,rud}$**	~~$N_{\delta,rud}$~~

CONTROL DERIVATIVES – JUST A TASTE

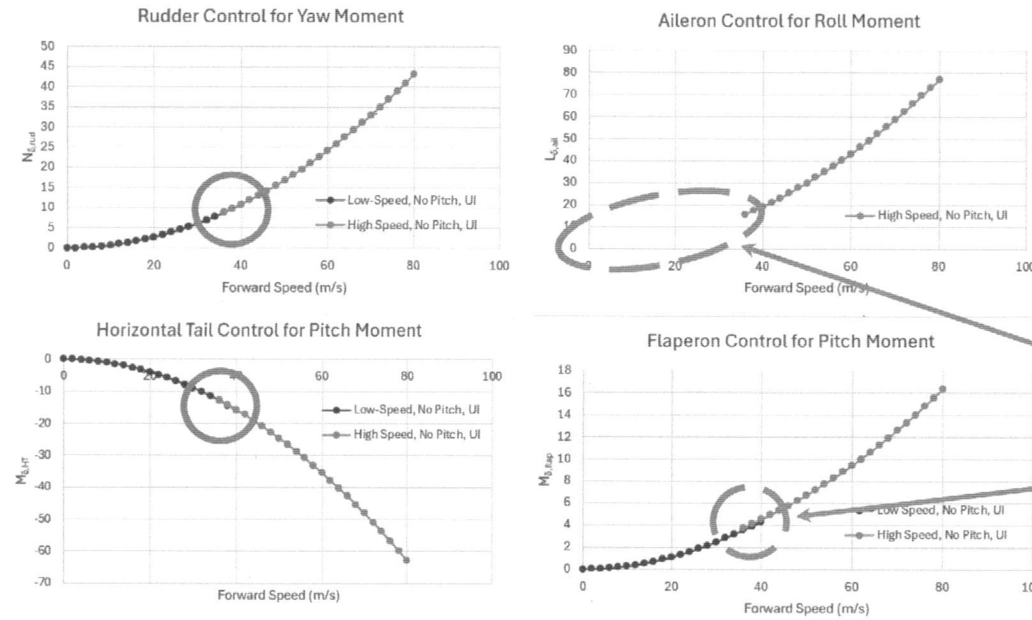

Other comparisons to consider:

$N_{\delta,ped}$ vs. $N_{\delta,rud}$ at low speed

$L_{\delta,lat}$ vs. $L_{\delta,ail}$ at low speed

Other RPM differential variables vs. their low speed control surface counterpart.

Flaperons are tied together at low speed and do not currently have a unique aileron derivative.

Flaperons separated at high speed, and do not have a unique flap derivative.

External Simulink controller should address RCAS limitations.

AGENDA

- Background & Motivation
- RCAS eVTOL Modeling
  - Model development
  - Trim strategies
- Stability & Control
- **Future Work**

eVTOLs: The Promise and the Risk

AOPA Pilot: https://www.aopa.org/news-and-media/all-news/2022/october/pilot/evtols-promise-and-risk

FUTURE WORK TO GET TO GOAL

- Simulink Model Development
- CONDUIT study with TDD/NASA Collaboration
- Over-actuated maneuvers study
  - –Hover sidestep
  - –Low-speed Turns
  - –Transition to Airplane Mode
- Linearized models from RCAS
  - –Interference
  - –Wake
  - –Etc.

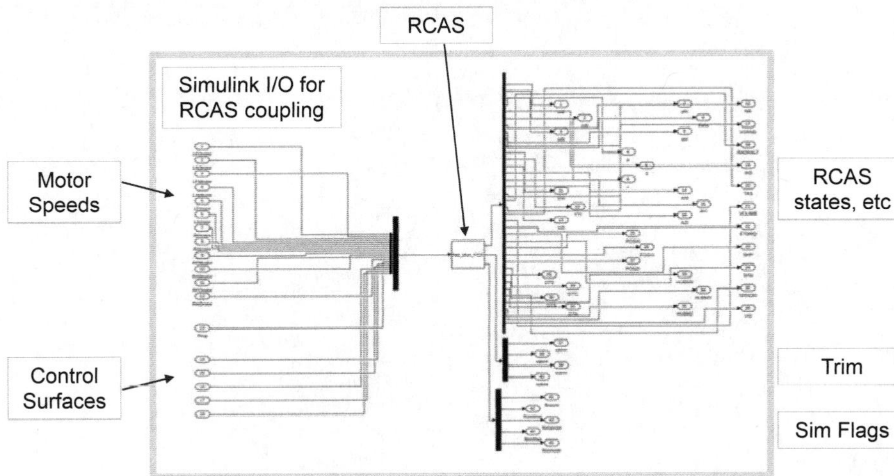

ACKNOWLEDGEMENTS

- Jean-Paul Reddinger, DEVCOM-ARL
- Laurent Binet, ONERA

DEVCOM
AVIATION & MISSILE CENTER

THANK YOU.

12 February 2025

25

U.S. ARMY

STI

Tiltrotor Flight Test Consulting LLC

Preliminary Findings From FAA Sponsored Flight Tests of Handling Qualities Task Elements for VTOL Aircraft

Martin Shubert
Shubertmw@gmail.com
Test Coordinator/Test Pilot
Tiltrotor Flight Test Consulting LLC

Mike Jones
mjones@systemstech.com
Test Director
Systems Technology Inc. (STI)
Hawthorne, CA, USA

1

The New Frontier

2

Background

Klyde, D. Mitchell, D., Shubert, M., Jones, M., *Handling Qualities Test Guide for Powered Lift VTOL Capable Aircraft with Indirect Flight Controls and Operating in Day Time Visual Flight Rules*, DOT/FAA/TC-23/59, March 2024 Final Report (referred to here as the Flight Test Guide)

- ➢ Introduction of Handling Qualities Task Elements (HQTE) concept along with catalogue of maneuvers.
- ➢ Toolbox of how to conduct time domain and frequency domain test inputs.

Follow-on effort to examine various aspects of assessment of HQ of VTOL aircraft with Indirect Flight Controls.

- ➢ Dedicated flight test campaign conducted to investigate the use of 'HQTEs' for civil certification

3

Flight Test Objectives

- In-flight FAA pilot evaluation of selected HQTE for appropriate civil performance tolerances with regards to:
  - Supporting the binary decision of certification as to whether the aircraft is airworthy and safe.
  - Stressing the design
  - Demonstrating agreement with predictive HQ metrics.
- In-flight FAA pilot evaluation of selected HQTE for course considerations to include: appropriate cueing, safety, and ease of setup/reconfiguration.
- Examination of tailoring of HQTE to provide insight into constraints and guidance for applicants in adapting the HQTE to their individual aircraft and concept of operations
- Illustration of the use of a flight test 'Scorecard' - relating supporting predictive flying qualities data to HQTE in meeting specified certification requirements.

4

Evaluation Aircraft

- Bell 412, C-FPGV, S/N 36034
- Operator: National Research Council of Canada
- Ottawa Macdonald-Cartier International Airport (CYOW)- NRC Flight Research Laboratory ADS-33 Course
- AW Certificate: Transport Canada Flight Permit: Experimental

Variable Stability System (VSS) that allows for tailoring of the fly-by-wire flight control design.

Aircraft Test Modifications

Changes to original Bell 412

- VSS-FBW and force feel system for Left Seat Pilot
- NRC Data Acquisition System with graphics display
- MCRUER System
  - Virtual Course display replaces Left Seat Pilot Displays
  - Provides secondary Data Acquisition System
  - Flight Engineer station for real-time monitoring.

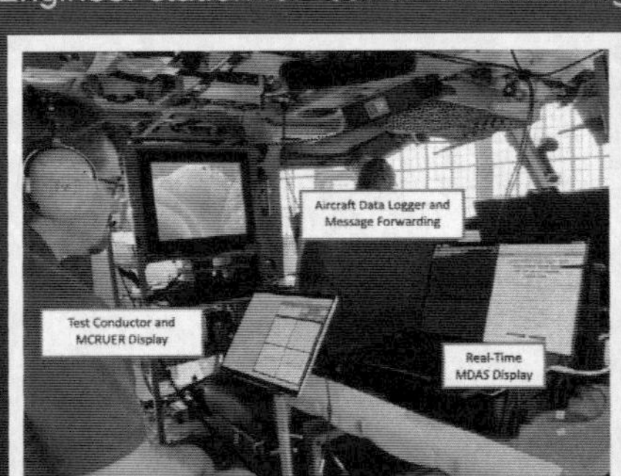

MCRUER Tablet System

Flight Test Execution

- 2 FAA pilots- 8 hours of flight test
- 5 (VTOL) HQTE were evaluated in various winds
  - ➢ Precision Hover
  - ➢ Hovering Turn and Hold
  - ➢ Pirouette
  - ➢ Lateral Reposition and Hold
  - ➢ Vertical Reposition and Hold (not fully evaluated due to time constraints)
- 3 planform response types- modeled through VSS
  - ➢ Level 1 Translational Rate Command
  - ➢ Level 1 Attitude Command Velocity Hold
  - ➢ Degraded Attitude Command Velocity Hold

Pretest Understanding of the HQTE Process

1. FAA goals for use in civil certification
2. Basic understanding of Application as a more quantitative approach to demonstrated HQ.
3. Use of Cooper Harper Rating (CHR) Scale to include definitions of performance, workload, and compensation
4. Current FAA/EASA position on adjudication of CHR
5. Current concerns of eVTOL applicants in the use of HQTE
6. Comparison of FTG, EUROCAE, and ADS-33 performance metrics

Suitability of Task Elements for Civil Cert.

- Representative of the operational elements of interest in the certification of the vehicle.
- Maneuvers potentially cover the areas of concern with advanced flight control concepts.
- Maneuvers are designed efficiently, with provisions for applicant tailoring.
- Maneuvers, when flown in combination, provide comprehensive coverage of the HQ elements in the low-speed environment which include:
  - Small amplitude precision maneuvers
  - Moderate amplitude aggressive maneuvers
  - Axes examined individually and in combination
  - Two speed regimes, near hover and higher groundspeeds
  - Calm winds and moderate winds

Ground Course Evaluation

- Safety- Plan on uncovering nasty issues and allow for safe task abandonment, consider alternative distances for target boards
- Cost- Minimize cost and allow for use of same course for multiple maneuvers
- Flexibility- ease of course setup and making the course work for various winds or reorientation for winds
- Differentiation in useable cueing from simulation to actual ground course.
  - Adjust maneuver in Draft Flight Test Guide which represents simulation
  - While ensuring that the simulation is a valuable tool for representing the actual task performance and workload
  - and can be used to decrease practice to proficiency

Precision Hover

Precision, Non-Aggressive

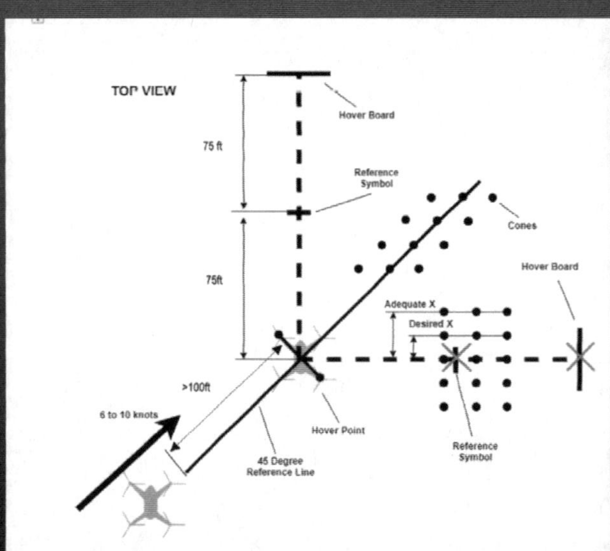

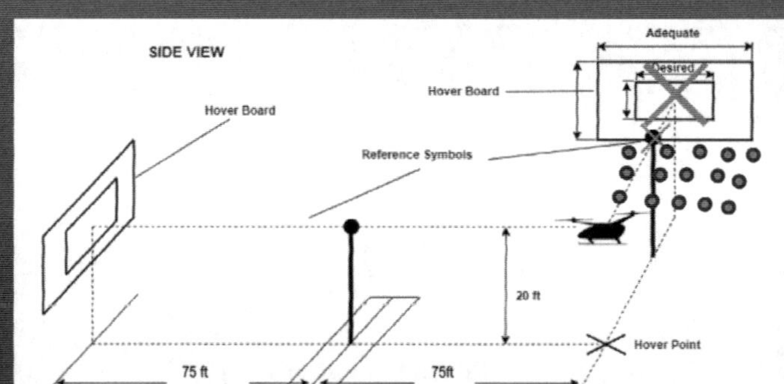

Flight Test Guide Constraints	Desired	Adequate
Attain a stabilized hover position from start of deceleration within:	5 sec	8 secs
Maintain lateral-longitudinal position within	± 3 ft	± 6 ft
Maintain altitude within:	± 2 ft	± 4 ft
Maintain heading within:	± 5 deg	± 10 deg
Maintain a stabilized hover for at least X seconds.	30 secs	30 secs
Residual motions from transition to hover or disturbances from stabilized hover	are easily damped with no oscillatory tendencies.	can be damped but are impacting performance. Any oscillatory behavior is not objectionable.

Precision Hover

ADS-33: Performance constraints similar to Cargo/Utility

Adequate time-10 sec
also Flown in mod Winds at most critical azimuth
"no objectionable oscillations in any axis either during the transition to hover or the stabilized hover"

TABLE A-1: TASK PERFORMANCE FTM VTOL HOVER

Performance	Desired	Adequate
Attain a stabilized hover within X seconds of initiation of deceleration	5	8
Maintain the longitudinal and lateral position within ±X meters from the ground-referenced hover point	± 0.10 D	± 0.20 D
Maintain the height within ±X feet	± 2	± 4
Maintain the heading within ±X degrees	± 5	± 10
Maintain stabilized hover for at least X seconds	10	10

ED 295, Assume D =33:

- greater position tolerances at 3.3ft and 6.6 ft
- shorter hover time
- Not flown in winds

FTG: also Flown in mod Winds at most critical azimuth

	Desired	Adequate
Attain a stabilized hover position from start of deceleration within:	5 sec	8 secs
Maintain lateral-longitudinal position within	± 3 ft	± 6 ft
Maintain altitude within:	± 2 ft	± 4 ft
Maintain heading within:	± 5 deg	± 10 deg
Maintain a stabilized hover for at least X seconds	30 secs	30 secs
Oscillations:	No objectionable oscillations in any axis either during the transition to hover or the stabilized hover.	No divergent oscillations in any axis either during the transition to hover or the stabilized hover.

Tailoring Options:

1. Shorter final hover time: 30 sec and 10 sec
2. Constrain inbound transition speed

Sample MCRUER Results

Precision Hover

Pilot 1: TRC flight at 6 kt and 10 kt run-in groundspeed

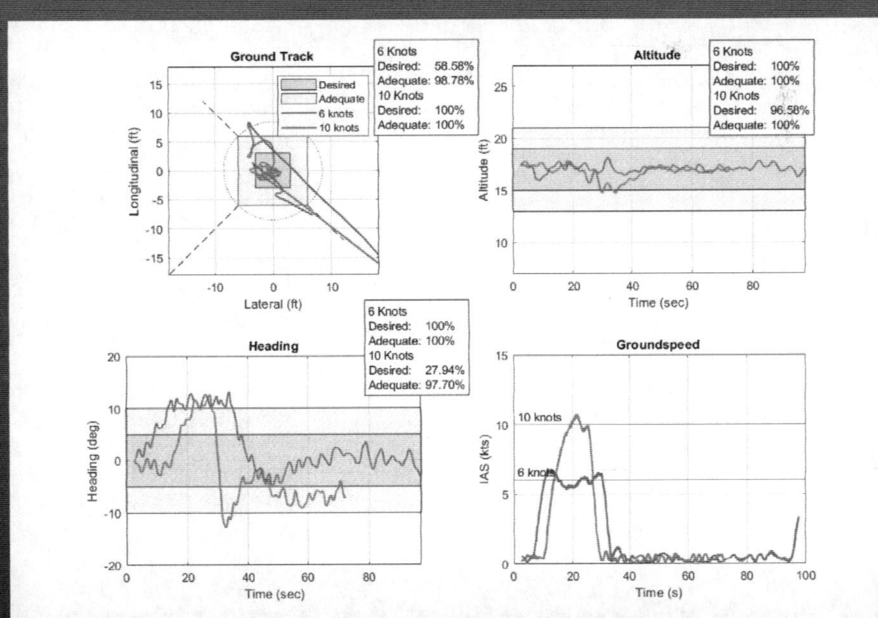

Precision Hover Findings

- Course:
  - sideboard should be replaced by cone-field.
  - lower forward boards provide for safer operations but if located too near the aircraft, can result in conflated longitudinal and vertical cueing issue

- The HQTE was suitable to civil certification with following recommended
  - ➢ Constrain speed capture for comment but do not include constraint in CHR for the maneuver. Run-in distance is not constrained
  - ➢ Evaluate multiple run-in speeds for state-based control response
  - ➢ Use 30 sec hover maintenance time
  - ➢ Wind constraint of critical azimuth is inappropriate, flight in moderate winds needs better definition

Hovering Turn and Hold

Precision, Non-Aggressive

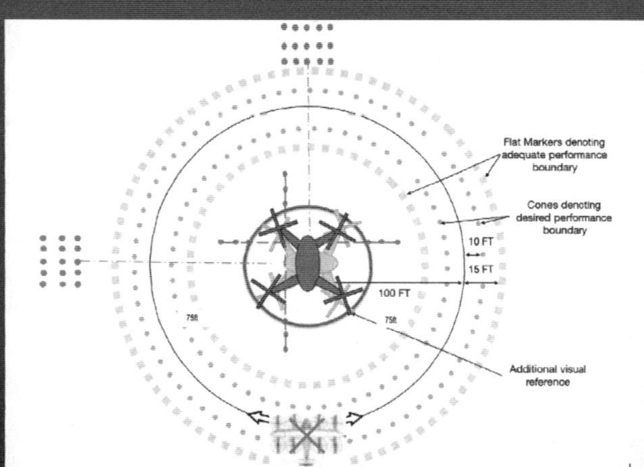

	Desired	Adequate
Maintain longitudinal and lateral position within ±X from the hover point.	±3 ft	±6 ft
Maintain altitude within ±X.	±3 ft	±6 ft
Stabilize the final rotorcraft heading at the 90 deg point and 270 deg point within ±X.	±5 deg	±10 deg
Complete maneuver within X.	≤ 50 s	≤ 60 s
Residual motions from transition to hover or disturbances from stabilized hover	*are easily damped with no oscillatory tendencies.*	*can be damped but are impacting performance. Any oscillatory behavior is not objectionable.*

Adjustments to Flight Test Guide Constraints

Hovering Turn and Hold Findings

- Course
  - Utilize Pirouette course for better cueing of drift during turn
  - Add two cone-fields 90 deg out from each other, and centered on Aircraft CG
  - Add two lines near aircraft pilot position for initial positioning to center CG
- The HQTE was suitable to civil certification with following recommended
  - Stay with 90 deg turn then 270 deg turn Tailoring was examined for shorter turns and for stopping at different azimuths but rejected
  - Choose time constraint to target a moderate turn rate of around 9 deg/sec to focus on precision vice aggression
  - Do not constrain setup orientation for moderate winds since full 360 deg turn is used and stopping position is 90 deg out from start position

General Observations

- The FAA's targeted performance metrics roughly agree with EUROCAE "D" factor of 33 ft in the ED-295 tables.
- The maneuver standards should not be scaled to either aircraft size or Flight Envelope boundary margins unless the boundaries are more restrictive than the maneuver descriptions.
- There is benefit to adding stabilized velocity captures for assessing advanced flight controls.
- Guidance is needed on maneuver symmetry arguments for reduced HQTE demonstrations.
- Guidance is needed on coaching and the use of display cueing.
- Guidance is needed on vertical setup to consider OGE/IGE.
- The MCRUER system was of benefit in course calibration and providing immediate feedback on pilot performance.

General Observations - Winds

- Testing in winds is recommended
- Adjustment of courses to wind direction is costly and inefficient
- Course layout and maneuver descriptions were modified to account for wind in any direction for moderate wind demonstration.
- Critical azimuth constraint was deemed inappropriate.
- Recommend moderate winds definition from 10-15 kts.

V-22 OGE
Wind Envelope

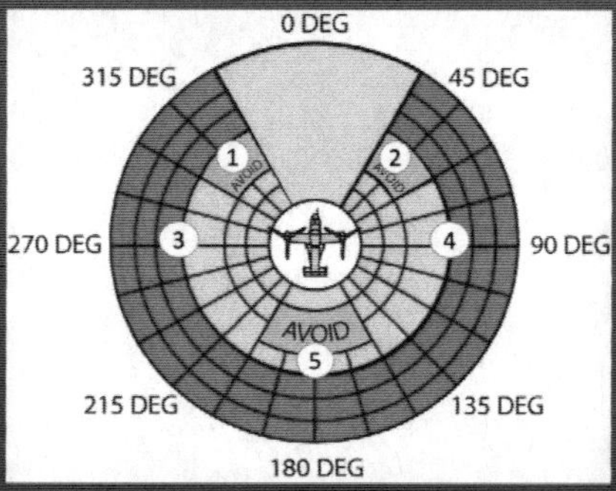

1&2- Pitch-up with Sideslip

3&4 - Asymmetric control power
- Increased power demand

5.- Lateral/directional Instability

General Observations- Oscillations

- Three of the maneuvers constraints for oscillations from FTG were found to be in conflict with both ADS-33 and EUROCAE guidance. MIL-DTL-32742(AR) does not have any mention in the maneuvers.
- Definition provided:

  PIO is a sustained or uncontrollable unintentional oscillation in which the airplane attitude, angular rate, normal acceleration, or other quantity derived from these states, is approximately 180 degrees out of phase with the pilot's control inputs.
- Nearly 180 out-of-phase should not be considered safe and airworthy.
- The HQTE requirements on objectional motions, if used, should be better described to agree with the PIO scale: to not require task abandonment to recover the aircraft (PIO rating from 1-3 acceptable)

Flight Test Guide Constraint	Desired	Adequate
Proposed Change	**Desired**	**Adequate**
Residual motions from transition to hover or disturbances from stabilized hover	are easily damped with no oscillatory tendencies.	can be damped but are impacting performance. Any oscillatory behavior is not objectionable.

Conclusion

- Tip of the Iceberg: Just a glimpse of work to come in finalizing HQTE as a Means of Compliance in certification of new evolving VTOL.
- This was a great opportunity to examine the processes for assessing Task Elements.
- Lessons learned here can also be applied toward improving Military specifications in the use of MTE.

Carl "Skin" Forsling

Penn '95

Aviation experience

- Marine Corps aviator: CH-46E, MV-22B
- Baltimore City Flight Officer: EC120
- Bell Textron BD: V-22, FMS, Advanced Programs
- Airbus US Space and Defense BD

Wife, grown kids, living in Alexandria, VA

Carl.Forsling@airbusus.com

Save Contact

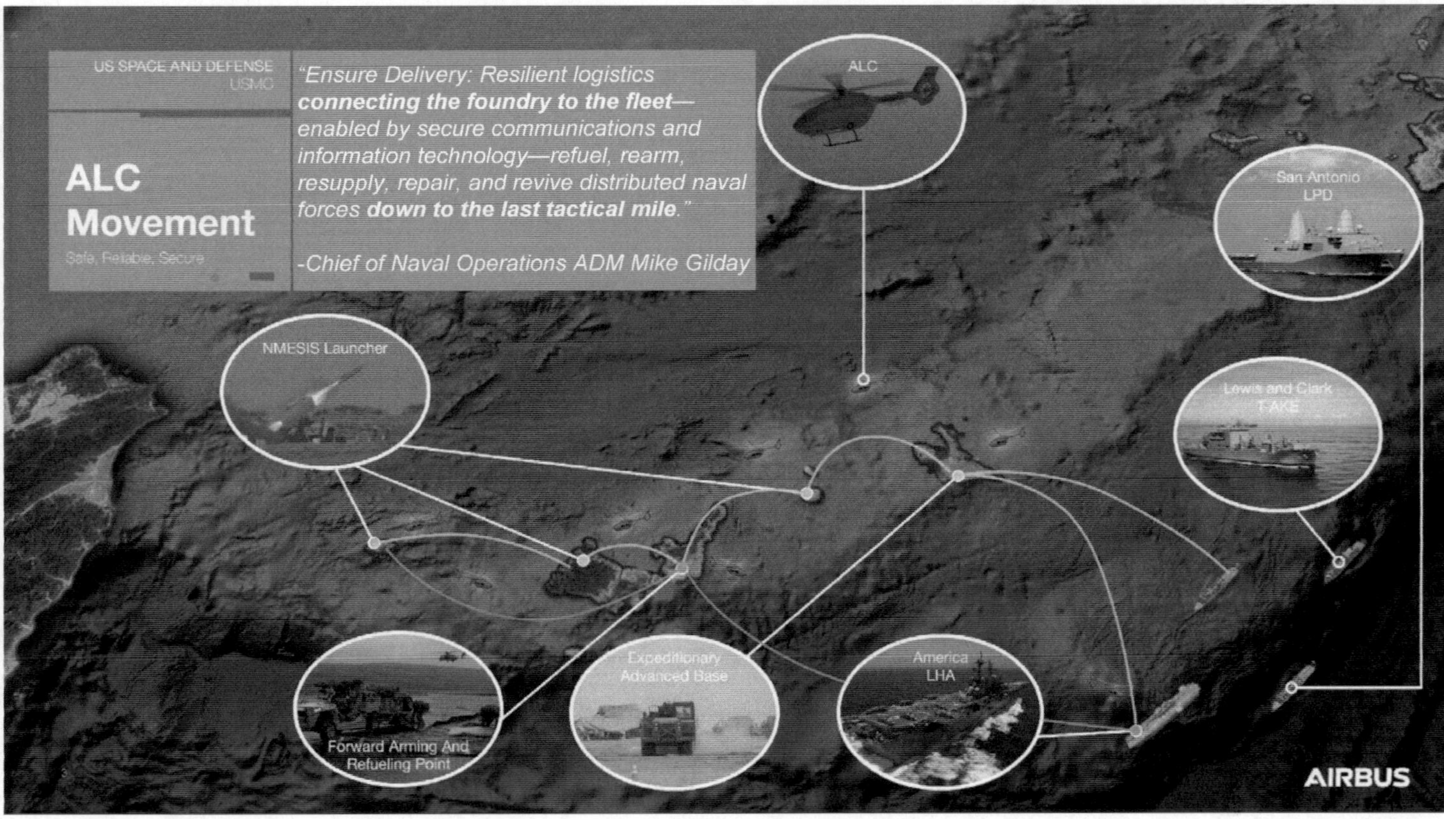

Aerial Logistics Connector

*The **Aerial Logistics Connector** is a NAVAIR program utilizing the MTA-RP pathway to develop an attritable autonomous cargo aircraft, designed to bridge the last tactical mile to logistical hubs. Four teams have been selected for Phase 1, with a down-select expected at the end of CY2025 or beginning of CY2026*

- Airbus US is proposing a variant of the UH-72B Lakota
- Working with top-tier partners to develop a fieldable prototype, aiming for scale production readiness by 2029
- The USMC's efforts add maritime resilience to a proven land-based aircraft, positioning it as an ideal platform for solving contested logistics challenges, from the plains of Europe to the corrosive environment of the South Pacific

OTAs and MTAs

Airbus Amber

Other Transaction Authority (OTA). Gives DoD flexibility to adopt and incorporate business practices from commercial industry standards and best practices.

Middle Tier of Acquisition (MTA). Enable rapid acquisition to deliver capability in 2-5 years with proven tech. Two pathways:

- *Rapid Prototyping.* Must field a prototype that can be demonstrated and provide a capability in an operational environment within 5 years of approved requirement
- *Rapid Fielding.* Fields proven technology. Must complete fielding within 5 years of approved requirement

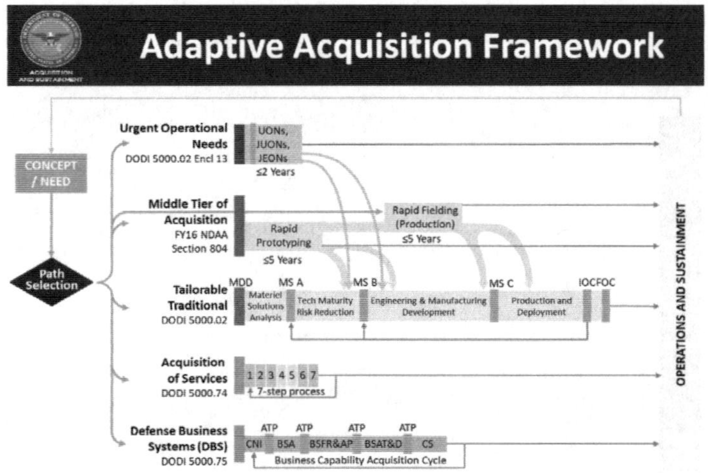

Middle Tier Acquisition – Rapid Prototyping

Purpose: *The rapid prototyping pathway provides for the use of innovative technologies* ***to rapidly develop fieldable prototypes*** *to demonstrate new capabilities and meet emerging military needs.*

- Highly adaptable to defense service needs.
- The illustration is provided to facilitate discussion; the actual schedule is determined by the government program.
- Ideal for **commercial-off-the-shelf** solutions.
- A **program transition plan is crucial**, as the DoD has multiple options due to this approach's flexibility.

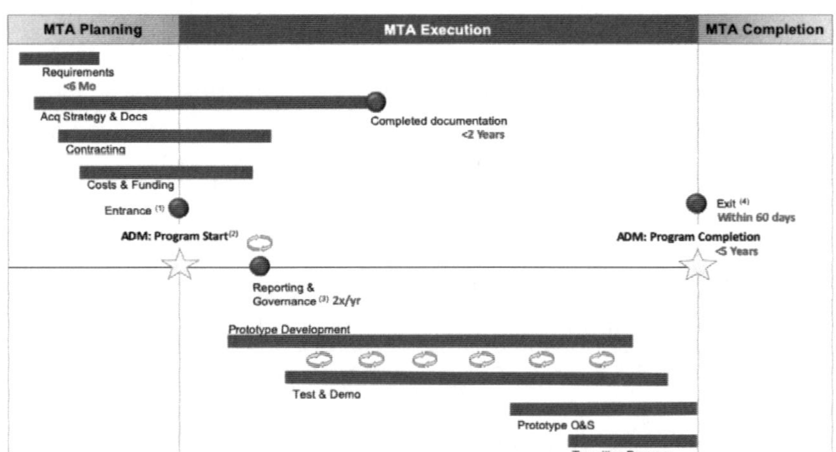

(1) *Major Systems*: Acquisition Decision Memorandum(ADM) signed by the Decision Authority (DA), Acquisition Strategy (which includes [1] Security, Schedule & Production Risks; [2] Test Strategy/Results; and [3] Transition Plan), and Program Identification Data (PID)
Non-Major Systems: ADM signed by the DA, PID
(2) Major Defense Acquisition Programs (MDAPs) require Under Secretary of Defense for Acquisition & Sustainment (USD(A&S)) Prior Written Approval
(3) Updated PID submitted twice a year with President's Budget and Program Objective Memorandum submissions to Office of Secretary of Defense (OSD)
(4) Signed Outcome ADM, Final PID, Assessment of Test Results

Program Schedule (Fiscal Year)

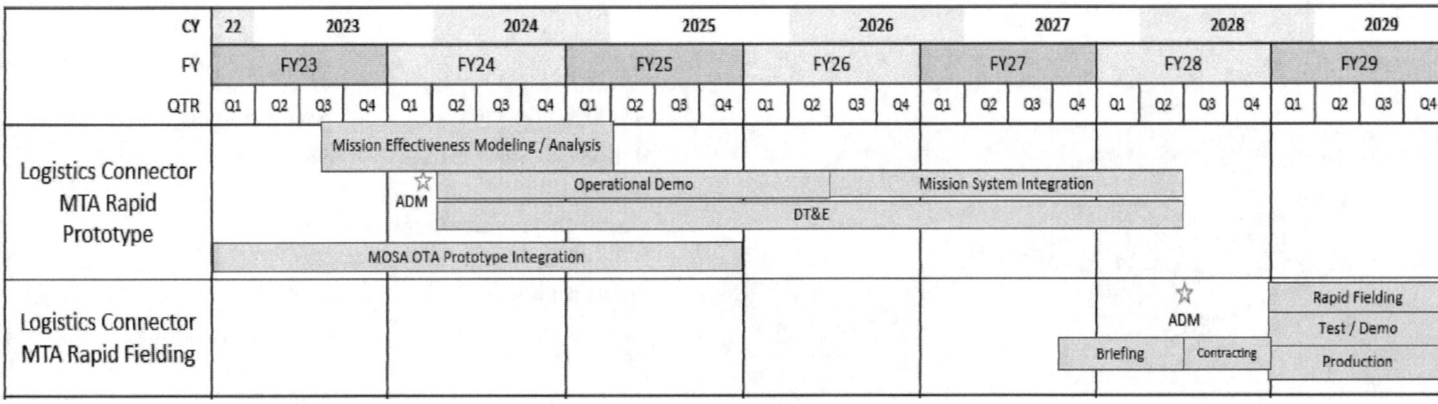

Page 7

AIRBUS

Evolution and Improvements

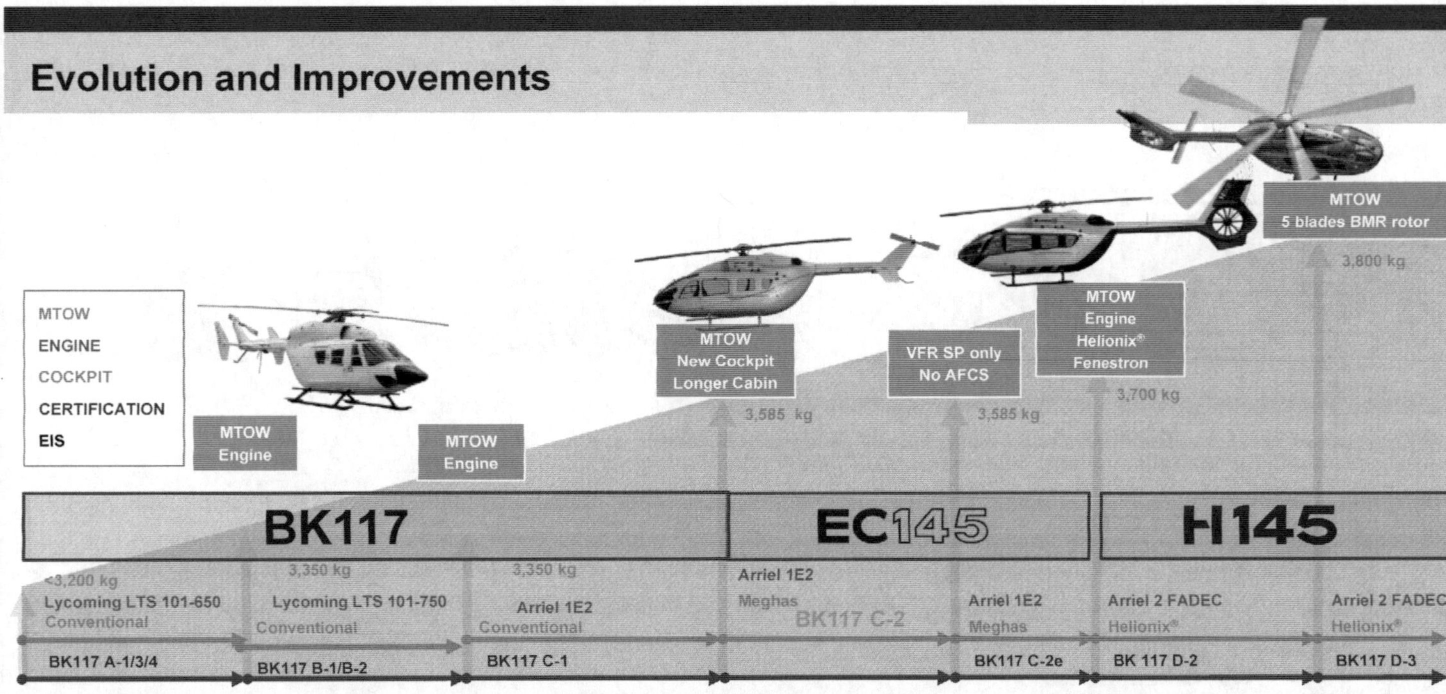

1617 H145s in service worldwide — 7.5M flight hours

H145D3/UH-72B Characteristics

Hover ceiling OGE	**8,955 ft**
Hover ceiling IGE	**Hover ceiling IGE**
Recommended cruise speed	**130 kts**
Max* range with standard fuel tanks at SL	**351 nm**
Rotor diameter	**35.43 ft**
Max takeoff weight (MTOW)	**8,378 lb**
Useful load	**4,200 lb**

479 UH-72A/B in Army and National Guard

9

H145 (UH-72B) Internal Cargo Capacity

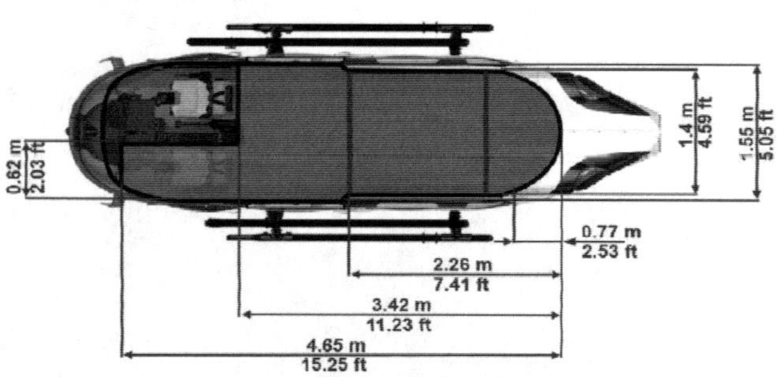

Floor Level

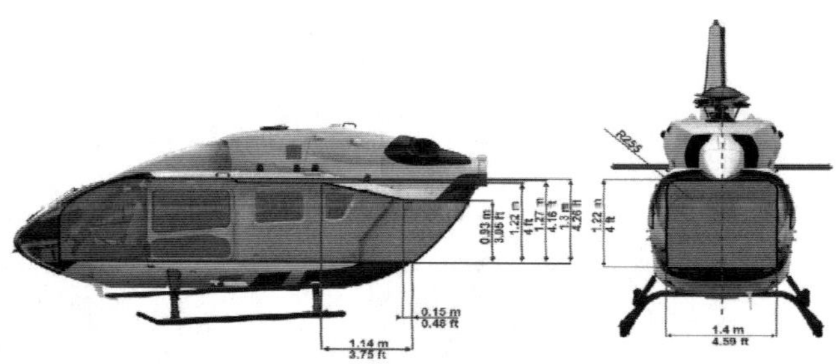

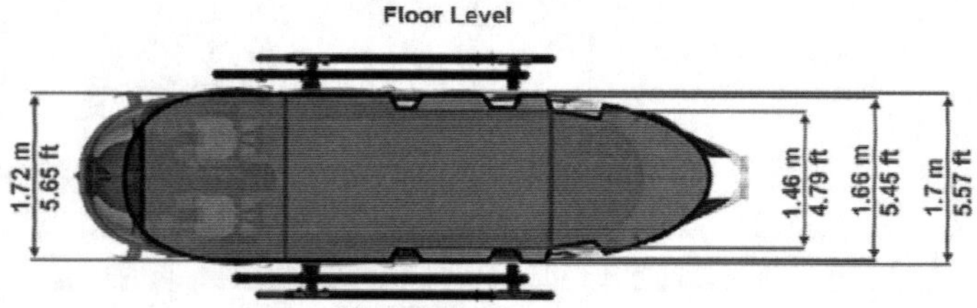

Shoulder Level

	Floor area		Volume	
Cabin & cargo compartment	4.72 m²	50.77 ft²	6.03 m³	212.83 ft³
(Cargo compartment with 8 seats installed)	0.84 m²	9.07 ft²	1.31 m³	46.40 ft³
Copilot	0.72 m²	7.73 ft²	0.80 m³	28.25 ft³
Pilot (not shaded)	1.09 m²	11.80 ft²	1.24 m³	43.76 ft³
Total (undivided)	5.43 m²	58.50 ft²	6.83 m³	241.08 ft³
Total (incl. pilot station)	6.52 m²	70.30 ft²	8.07 m³	284.84 ft³

10

H145 (UH-72B) Internal Cargo Capacity

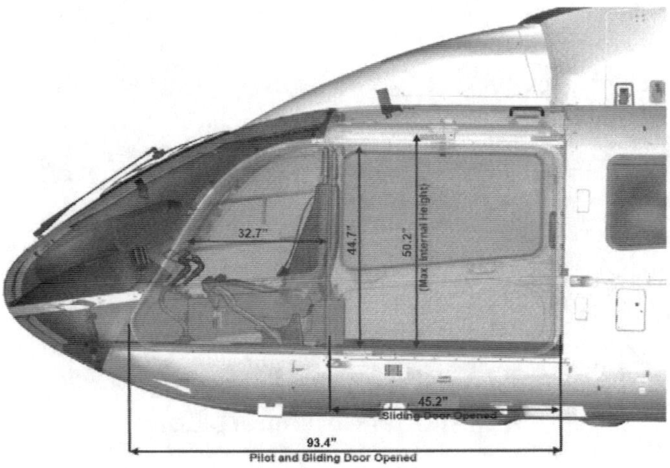

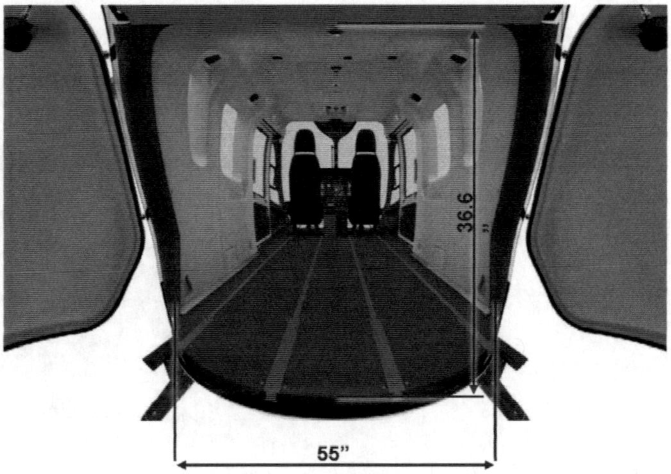

11

Key Differences

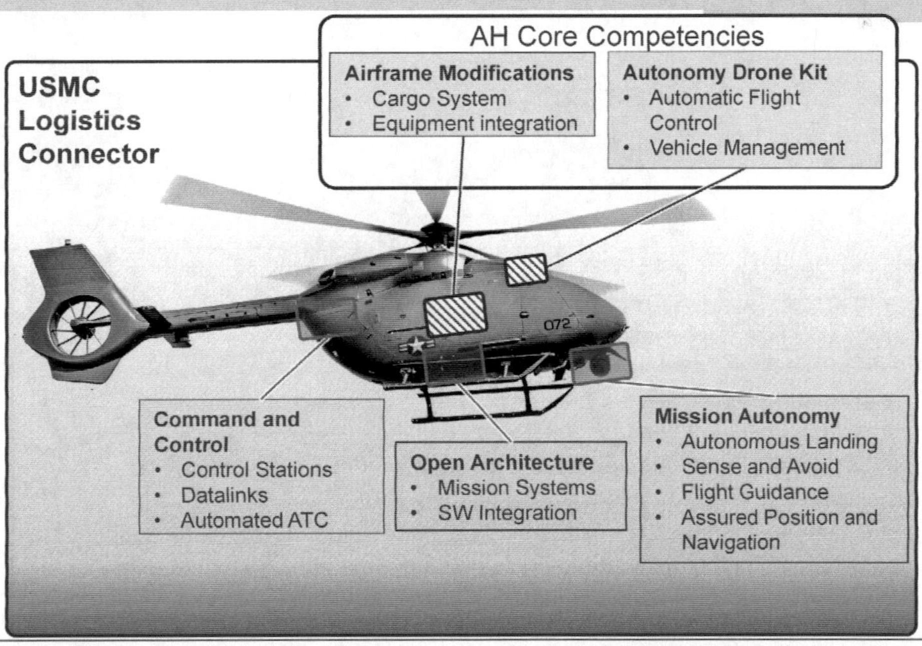

USMC Load Requirements

**** NAVAIR requirement it 2x JMICs. Airbus design can hold 3x JMICs***

Nose Door Design (Ongoing)

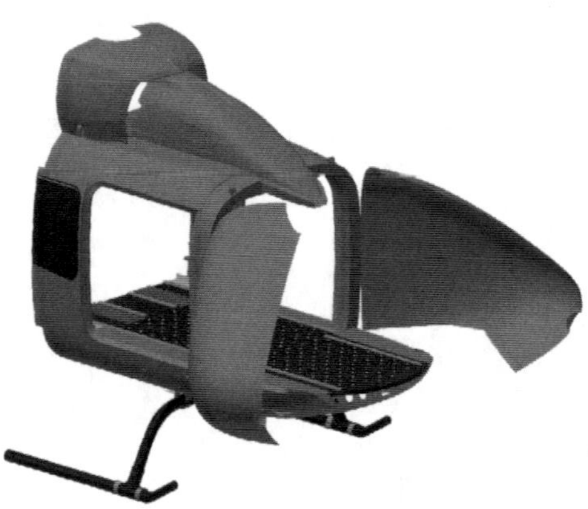

Clamshell Door

Side Swing Door

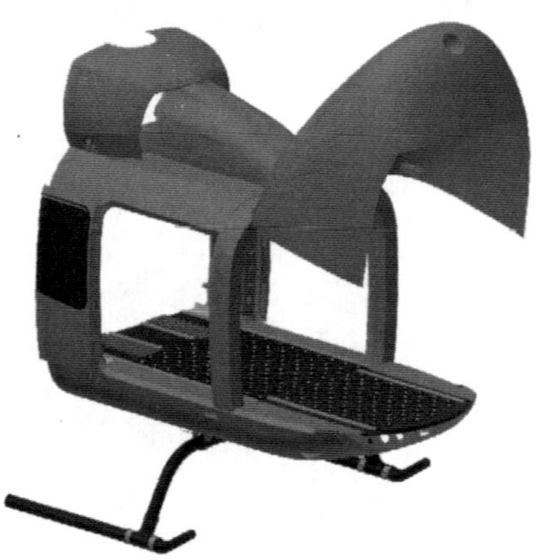

Upswing Door

Cargo mockup

15

This document contains proprietary data or other confidential information of Airbus Helicopters, Inc. ("AHI") or its affiliates, which shall not be disclosed to any third party or reproduced, in whole or in part, without AHI's prior written consent.

This document does not contain export-controlled information as defined by the International Traffic in Arms Regulations (ITAR) or the Export Administration Regulations (EAR)

AIRBUS

Export control
Not technical

HELICOPTERS

Questions

"This document does not contain export-controlled information as defined by the International Traffic in Arms Regulations (ITAR) or the Export Administration Regulations (EAR)"

AIRBUS

Evaluation of Active Sidestick Failure Mode Handling Qualities

Tom Berger
U.S. Army Combat Capabilities Development
Command Aviation & Missile Center
Moffett Field, CA, USA

Kevin Christensen
Bell Textron Inc.
Fort Worth, Texas, USA

Seung Baik
Bell Textron Inc.
Fort Worth, Texas, USA

John (Jack) Rominger
Bell Textron Inc.
Fort Worth, Texas, USA

Sam Clark
IronMountain Solutions
Future Long Range Assault Aircraft PM
Huntsville, AL, USA

ABSTRACT

A piloted simulation experiment was conducted in the NASA Ames Vertical Motion Simulator to investigate the handling qualities degradation as a result of an active sidestick failure mode. The particular failure mode investigated in this effort was loss of active force feel resulting in a friction only force feel (no force gradient or detent to center the stick to the neutral position). Three different friction levels were evaluated (Light 0.4 lbf, Medium 0.9 lbf, and Heavy 1.4 lbf). The evaluation used a generic utility-class tiltrotor model. Five Mission Task Elements and one operational scenario were used, with evaluations being performed by three experimental test pilots. Overall, Level 2 handling qualities were achievable for the three failure mode configurations, and no Pilot Induced Oscillations were encountered. Additionally, pilots were able to continue high-speed terrain flight and accomplish a transition to hover with all three failure mode configurations during the operational scenario evaluation.

INTRODUCTION

With the move towards fly-by-wire flight control systems for rotorcraft, pilot inceptors are no longer physically connected to the mechanical hardware of the aircraft. This has allowed the move to smaller, lighter, side-mounted (sidestick) controllers, which are typically unique trim and can be either passive or active. Passive sidesticks use a mechanical spring and damper system to provide force-feel characteristics, while active sidesticks typically include motors, clutches, position and force sensors, and closed-loop control to provide force-feel characteristics. There are many benefits to active sidesticks including configurable force-feel characteristics based on the flight condition or response type, tactile cueing, and electronic linking between pilot and copilot inceptors, which is important for pilot awareness (Ref. 1). These benefits, though, come at the cost of potential active sidestick failure modes due to the increased complexity over passive sidesticks.

While active sidesticks will be designed to meet reliability requirements, the SAE International active sidestick recommended practice SAE ARP5764 (Ref. 2) specifies that active inceptors should be designed such that they can be used in the event of a loss of active feel force mode. In particular, SAE ARP5764 states that in a failure mode: "some degradation of inceptor feel qualities is acceptable, however, the pilot's ability to fly the aircraft should not be impaired." This is consistent with MIL-DTL-32742 (Ref. 3) which specifies that at least Level 2 handling qualities should be achievable with failure modes of certain probability.

The failure mode investigated in this study is loss of active force feel resulting in a friction only force feel. Such a failure mode has been previously addressed in the literature (no force gradient or detent to center the stick to the neutral position). In particular, Ref. 4, which discusses performance requirements for active inceptors, states that in a loss of force feel failure mode, the desired friction value is a trade-off between reasonably light friction forces for ease of moving the stick and high enough friction force to prevent feedback of aircraft accelerations through the controls, although specific friction values are not listed. In Ref. 5, a generic test for transient recovery of an active inceptor system failure to a friction only condition was developed and implemented in a motion-based simulator. One friction value was evaluated in that work, based on the mechanical characteristics of the inceptor used, and the results showed that the transient recovery from the failure often resulted in over-controlling, pilot induced oscillations (PIO), or unintentional control inputs. Reference 5 did not look at the long-term handling qualities of the failure mode, but suggested they should be evaluated using standard handling qualities methods (Mission Task Elements, MTEs).

Presented at the Vertical Flight Society's 2025 Handling Qualities Technical Meeting, Huntsville, AL, January 22–23, 2025.

MIL-DTL-32742 provides minimum force gradients for Level 1 and 2 handling qualities. The requirement specifies that a force gradient of at least a 0.5 lbf/in is required in both the lateral and longitudinal axes for Level 2 handling qualities, which suggests that Level 2 handling qualities cannot be attained with friction only and no force gradient. However, the force gradient requirement in MIL-DTL-32742, which is carried over from MIL-F-83300 (Ref. 6), is primarily based on centerstick configurations and existing Vertical/Short Takeoff and Landing (V/STOL) aircraft at the time MIL-F-83300 was published (1970).

To determine if Level 2 or better handling qualities can be attained with a friction only force feel sidestick, and what values of friction are too light and too heavy, a piloted handling qualities simulation evaluation was conducted and described in this paper. The simulation evaluation was carried out in the NASA Ames Vertical Motion Simulator, a large range of motion simulator to account for any feedback of aircraft accelerations through the controls. The evaluation used the U.S. Army Combat Capability Development Command (DEVCOM) Aviation & Missile Center (AvMC) generic tiltrotor flight dynamics model and flight control system. Five MTEs, representative of tasks pilots may need to fly with active sidestick failures, and one operational scenario were used. One Army and two Bell experimental test pilots participated in the evaluation.

The remainder of the paper covers a brief description of the tiltrotor model and control laws. Next, details of the sidestick failure mode are provided. An overview of the simulation experiment is given, including descriptions of the tasks flown. Finally, the results of the simulation experiment are provided, along with conclusions.

VEHICLE DESCRIPTION

Bare-Airframe Model

A schematic of the vehicle used in this study is shown in Fig. 1. The flight-dynamics model of this generic utility-class tiltrotor configuration was developed using HeliUM-A, DEVCOM AvMC's in-house flight-dynamics modeling software tool developed as an extension to the University of Maryland's HeliUM simulation model (Refs. 7, 8). HeliUM-A uses a finite-element approach to model flexible rotor blades with coupled non-linear flap/lag/torsion dynamics to capture structural, inertial, and aerodynamic loads along each blade segment. Blade, wing, and fuselage aerodynamics come from non-linear lookup tables, and the rotor airwakes are modeled using a dynamic inflow model.

The model is meant to represent a generic configuration and not a specific industry design. The aircraft has a gross weight of 32,000 lbs and falls into the Future Vertical Lift (FVL) Capabilities Set #3 category. The configuration was derived from scaling geometric, inertial, and structural properties of the XV-15, V-22, and the notional NASA Large Civil Tilt-Rotor 2 (LCTR2). Reference 9 presents a detailed description of the model.

Fig. 1. Generic utility-class tiltrotor schematics.

Linear state-space point models and trim data were extracted from HeliUM-A at a range of airspeeds and nacelle tilt angles. The point models were used to develop the flight control system. In addition, the point models and trim data were combined to form a continuous full flight envelope quasi-linear parameter varying (qLPV) stitched simulation model (Ref. 10). The model is suitable for real-time simulation, and formed the basis of the simulation model used in the experiment described here.

Control Laws

Full flight envelope control laws were previously developed for this generic tiltrotor and tested in piloted simulation (Ref. 11). An explicit model following architecture was used and optimized in CONDUIT® (Ref. 12) to meet a comprehensive set of stability, handling qualities, and performance specifications.

The control laws provide pilot selectable response types and hold modes. This is mechanized in three Control Law (CL) modes—CL0 through CL2—with increasing levels of augmentation, and an additional high-speed velocity hold mode. Figure 2 shows the response types and hold modes in each axis for each CL mode. Pilots can toggle between the control law modes ("upshift" to higher levels of augmentation or "downshift" to lower levels) using a Mode Toggle Switch on the sidestick.

Control Law Mode 0 (CL0) provides the minimum level of augmentation and maximum level of agility, and is intended for use in Good Visual Environment (GVE). CL0 provides a Rate Command/Attitude Hold (RCAH) response type in the roll axis for all airspeeds, and RCAH in the pitch axis from hover to 200 kts. Above 200 kts, the pitch axis response type transitions to a G (normal acceleration) Command/Alpha (angle of attack) Hold (GCAH). The thrust control lever (TCL) commands rotor torque. The pedals provide a Rate Command/Direction Hold (RCDH) response type from hover to 25 kts, then Rate Command with automatic Turn Coordination (RCTC) from 25 to 100 kts, and Beta (sideslip angle) Command with automatic Turn Coordination (BCTC) above 100 kts. Turn coordination allows pilots to fly feet on the floor in forward flight.

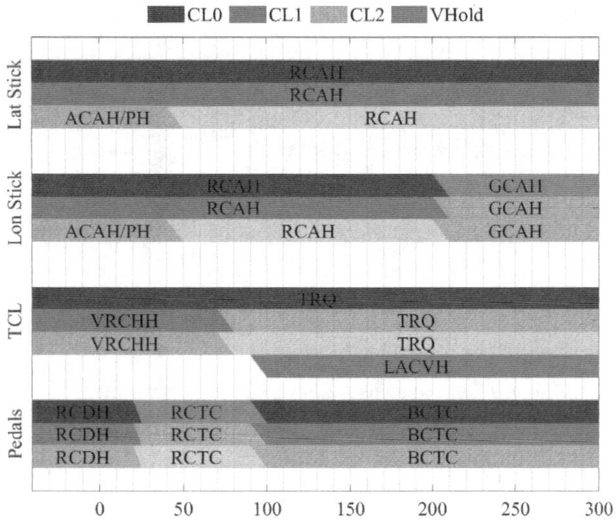

- RCAH = Rate Command/Attitude Hold
- ACAH/PH = Attitude Command/Attitude Hold with Position Hold
- GCAH = G (normal acceleration) Command/Alpha (angle of attack) Hold
- TRQ = Torque Command
- VRCHH = Vertical Rate Command/Height Hold
- LACVH = Linear Acceleration Command/Velocity Hold
- RCDH = Rate Command/Direction Hold
- RCTC = Rate Command with automatic Turn Coordination
- BCTC = Beta (sideslip angle) Command with automatic Turn Coordination

Fig. 2. Control law response types and hold modes.

Control Law Mode 1 (CL1) provides the same response types and hold modes as CL0, with the addition of Vertical Rate Command with Height Hold (VRCHH) in the thrust axis from hover to 60 kts. Above 60 kts, the TCL commands rotor torque.

Control Law Mode 2 (CL2) provides an Attitude Command/Attitude Hold (ACAH) response type in the pitch and roll axes from hover to 40 kts. In addition, Position Hold (PH) is automatically armed in CL2. Position Hold engages when the stick is in detent and ground speed is below GS = 3 kts. At that point, the control laws engage an automatic deceleration to hover and latch on to a position when ground speed drops below GS = 0.5 kts.

Mode annunciation and reference information is provided on the panel mounted displays.

For this test, evaluations were done at hover (using CL1 for GVE and CL2 for DVE) and at 200 kts with Velocity Hold engaged. Figure 3(a) shows the pitch and roll axis attitude bandwidth and phase delay in hover for the RCAH (CL1) and ACAH (CL2) response types. Figure 3(b) shows the pitch and roll axis attitude bandwidth and phase delay at 200 kts. Note that the visual and motion system delay of the VMS (roughly equal to 0.058 sec) was added to the attitude frequency responses to determine the bandwidth and phase delay, thus matching the visual and motion cues provided to the pilots during the experiment. All configurations tested meet the Level 1 attitude bandwidth and phase delay requirements. Table 1 lists the achievable rates and attitudes in hover and at 200 kts in the pitch and roll axes.

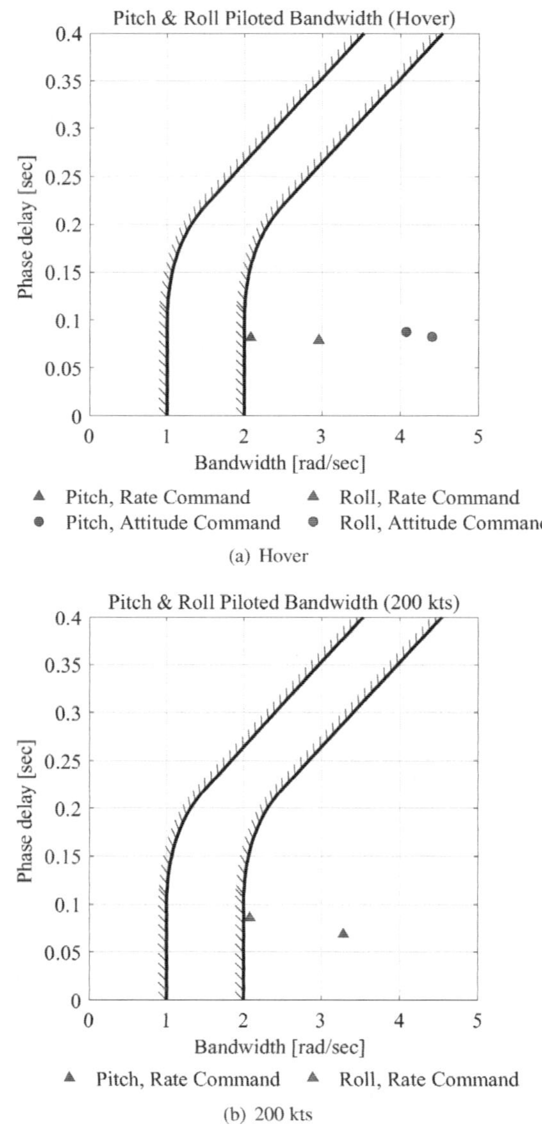

Fig. 3. Pitch and roll axis attitude bandwidth and phase delay

Inceptor Characteristics

Table 2 lists the sidestick force-feel characteristics for the active and three failure mode configurations tested. In the Active configuration, the lateral axis has a steeper gradient for left (inward) versus right (outward) arm motions. The longitudi-

Table 1. Achievable Rates and Attitudes

	Hover	200 kts
Rate-Command		
Pitch	30 deg/sec	18 deg/sec
Roll	45 deg/sec	60 deg/sec
Attitude-Command		
Pitch	30 deg	–
Roll	60 deg	–

nal axis has a symmetric force gradient that is scheduled with airspeed, with a heavier gradient at high speed. The failure mode configurations have zero preload and gradient values, thus providing no stick re-centering or detent, and friction values of 0.4 lbf (Light), 0.9 lbf (Medium), and 1.4 lbf (Heavy).

SIMULATION OVERVIEW

Simulation Facilities

The experiment was conducted in the NASA Ames Vertical Motion Simulator (VMS) which provides 6-degree of freedom motion with 60 feet of vertical, 40 feet of lateral, and 8 feet of longitudinal travel. Figure 4(a) shows a schematic of the simulator facility. The side-by-side rotorcraft cockpit know as T-cab was used, shown in Fig. 4(b). Evaluations were flown from the right seat, which was equipped with an electro-hydraulic McFadden sidestick inceptor (Ref. 13), thrust control lever (TCL), and pedals.

Figure 5 shows a force versus displacement plot for the lateral axis for the Active and three failure mode configurations. The plot shows the measured force values as well as the commanded values. For all configurations, there is an excellent agreement between the measured and commanded values. Similarly good matches between the measured and commanded force-displacement curves was seen in the longitudinal axis.

Evaluation Tasks

Mission Task Elements Five MTEs were used in the evaluation, selected to cover tasks that pilots may need to accomplish with sidestick failures over a range of flight conditions. The MTEs used were:

1. Hover, Good Visual Environment (hover)
2. Hover, Degraded Visual Environment (hover)
3. Pitch Attitude Capture and Hold (200 kts)
4. Bank Angle Capture and Hold (200 kts)
5. Station Keeping (200 kts)

The Hover MTEs were flown using the description and standards provided in MIL-DTL-32742. Degraded Visual Environment (DVE) was simulated using fog in the out the window visual scene. The fog Runway Visual Range (RVR) was set to

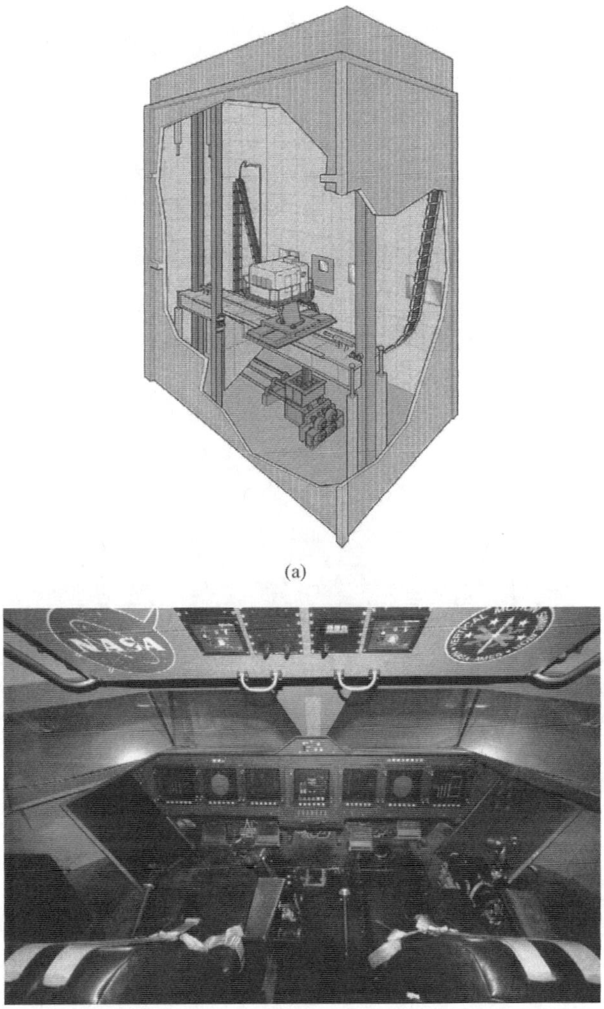
(a)

(b)

Fig. 4. NASA Ames Vertical Motion Simulator

500 ft when the aircraft was more than 400 ft away from the hover point (so that the pilot could see the hover boards from the staring point of the MTE) and 250 ft when the aircraft was less than 6 ft from the hover point, with a linear fade in between. Figure 6 shows the forward out the window view from the hover point.

The Pitch and Bank Angle Capture and Hold MTEs were flown at 200 kts using the description and standards provided in Ref. 14. The Station Keeping MTE was also flown at 200 kts, and is described here.

The objectives of the Station Keeping MTE are:

- Check for handling qualities while station keeping off another aircraft.
- Check for harmony between pitch, roll, and thrust controllers.

The Station Keeping MTE task description is:

Table 2. Inceptor Characteristics

	Active		Failure Modes*		
	Lateral	Longitudinal	Light	Medium	Heavy
Displacement (in)	±1.7	±1.7	±1.7	±1.7	±1.7
Gradient (lbf/in)	2.21 (left) / 1.60 (right)	2.94 (low-speed) / 5.87 (high-speed)	0	0	0
Preload (lbf)	0.4	0.4	0	0	0
Friction (lbf)	0.2	0.2	0.4	0.9	1.4

* Applies to lateral and longitudinal axes

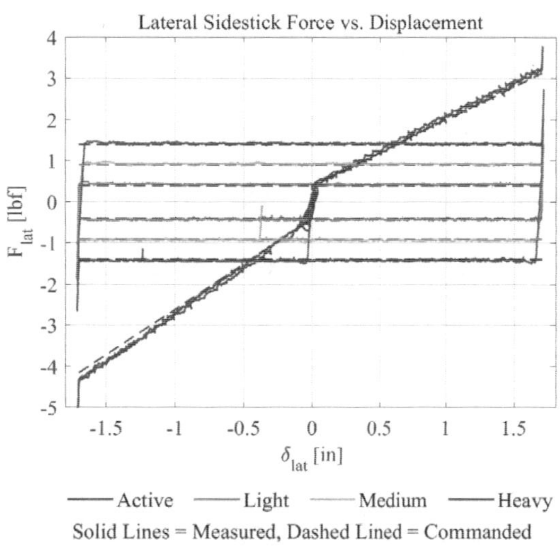

Fig. 5. Lateral sidestick force vs displacement for the active and friction only modes.

- Start in the observation position approximately 200 ft away from the tanker aircraft on a 45 deg diagonal.
- Establish and hold simulated refueling position for 30 sec (shown in Fig. 7).

The performance criteria of the Station Keeping MTE are:

- Stabilize in simulated refueling position within 30 sec (desired) / 40 sec (adequate).
- Maintain lateral simulated refueling position between tanker aircraft wingtip and outboard pylon for 100% time (desired) / 50% time (adequate).
- Maintain vertical simulated refueling position keeping tanker aircraft horizontal tail between wing trailing edge and top of propellers for 100% time (desired) / 50% time (adequate).
- Maintain fore/aft simulated refueling position keeping tanker aircraft opposite horizontal tail tip between wingtip and outboard pylon for 100% time (desired) / 50% time (adequate).

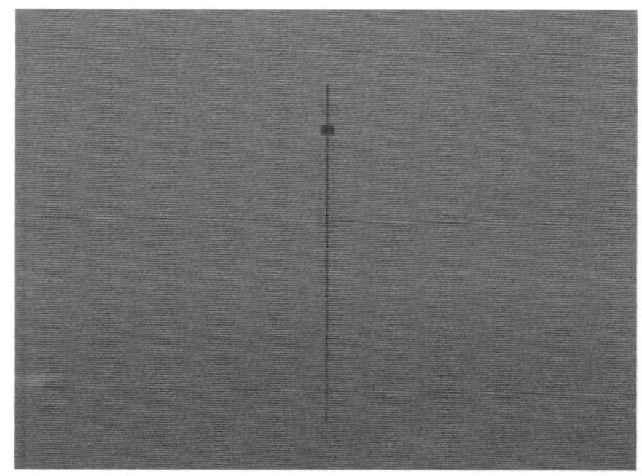

Fig. 6. Hover MTE Degraded Visual Environment (DVE) forward out the window view in the hover position.

Fig. 7. Station Keeping MTE simulated refueling position.

Operational Scenario In addition to the MTEs described above, which are used to assess handling qualities of the failure mode, a simulated operational scenario consisting of low-level, high-speed flight through rolling terrain was also used. The operational scenario was flown in the Camp Dawson, WV area, and is shown in Fig. 8.

Pilots began flying the scenario at 240 kts and were required to stay within a defined corridor (shown in Fig. 8). The length of the route flown was approximately 20 nm. Pilots were required to fly low and use the terrain for masking to avoid detection by generic radar threats throughout the database. At the end of the scenario, pilots converted from cruise/airplane mode down to helicopter mode and ended in a stable hover over a designated point.

Pilots began flying the scenario with the Active sidestick configuration, and the failure mode was injected at a random point during the scenario. In additional to evaluating the failure transient, the scenario also evaluated the ability to continue high-speed terrain flight and convert from high-speed flight to hover with a failed sidestick.

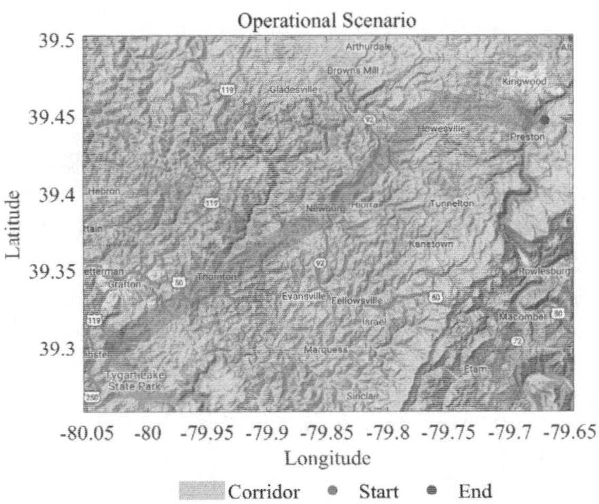

Fig. 8. Operational scenario map view.

Pilot Demographics and Rating Scales

Three experimental test pilots participated in the study, two Bell pilots and one Army pilot. The pilot experience ranged from 3000 to 5000 flight hours (with an average of 4000 hrs) primarily in H-60 and H-47 variants, as well as some experience in the V-22 and V-280 Technology Demonstrator.

After evaluating each configuration using a particular MTE, pilots provided comments as well as Cooper-Harper Handling Qualities Ratings (HQR) and PIO Ratings (PIOR).

RESULTS

Hover MTE (GVE)

Figure 9 shows the average (markers) as well as best and worst (error bars) HQRs and PIORs for the Hover MTE tested in GVE for the Active and three failure mode sidestick configurations. In GVE, pilots flew this task using control law mode CL1, which provided a RCAH response type in pitch and roll, VRCHH in the heave axis, and RCDH in the yaw axis, as shown in Fig. 2.

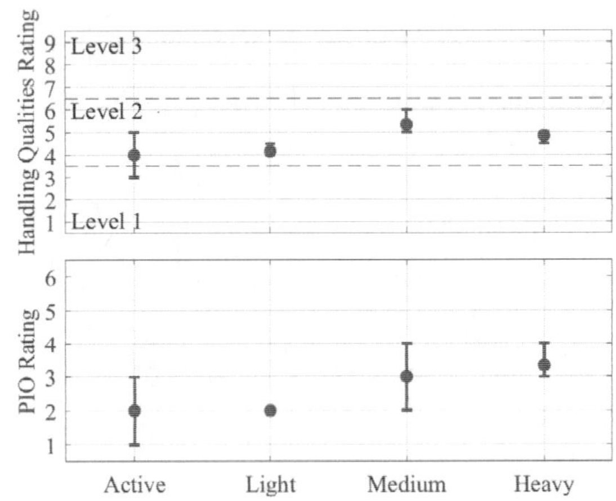

Fig. 9. Handling Qualities and Pilot Induced Oscillation Ratings for the Hover MTE tested in GVE. (Markers indicate average ratings and error bars indicate best and worst ratings.)

Figure 10 shows the aircraft position during the 30-sec hover portion of the task, along with the desired and adequate performance bounds. Overall, desired performance was attainable with the Active configuration, which received average HQR 4.0, driven by the required pilot compensation, and PIOR 2.0, suggesting that task performance was not compromised by oscillations or undesirable motions.

Desired performance was also attainable with the Light friction configuration, which received average HQR 4.2 and PIOR 2.0. For the Light friction configuration, pilots commented that they could "fine tune their inputs" and "maintain better performance as compared to the Medium and Heavy failure configurations." In addition, pilots commented that the Light friction was "tolerable and preferred over the other friction values for this task."

For the Medium friction configuration, overall adequate performance was achievable, and the configuration was rated an average HQR 5.3 and PIOR 3.0, suggesting considerable to extensive pilot compensation to achieve adequate performance and undesirable motions that compromised task performance. For this configuration, pilots commented that it was "undesirable but predictable" and felt "sluggish." Pilots "never felt...out of phase" with oscillations and further commented that this configuration was better than the Heavy friction configuration and that the "friction [was] light enough that it is acceptable for the task."

For the Heavy friction configuration, overall adequate performance was achievable. The Heavy configuration was rated a slightly better average HQR (4.8) and slightly worse average PIOR (3.3) than the Medium configuration. For this configuration, pilots commented that the "increased forces made it feel sluggish" and that they thought the "friction [was] much too high."

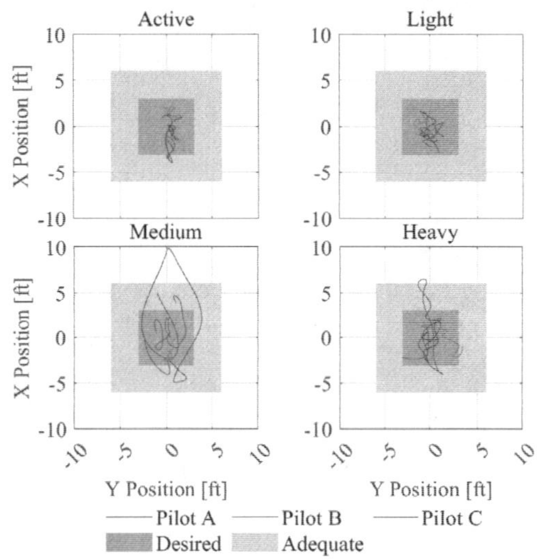

Fig. 10. Position hold performance during 30-sec hover capture portion of the Hover MTE (GVE).

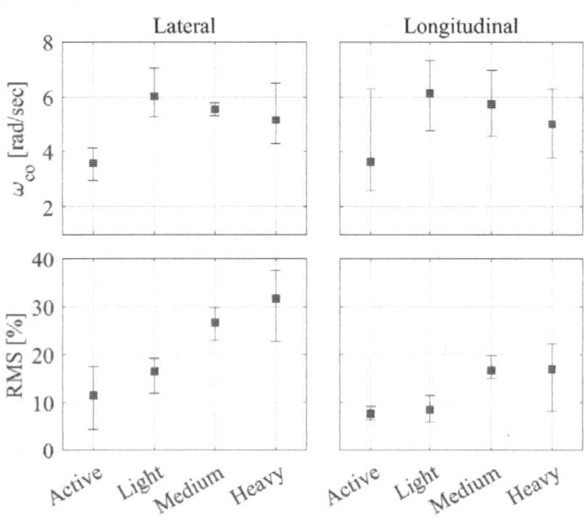

Fig. 11. Pilot force input cutoff frequency and RMS (Hover MTE, GVE).

Figure 11 shows the pilot cutoff frequency ω_{co} in the lateral and longitudinal axes as well as the stick root mean square (RMS) in both axes, for pilot stick force F inputs, during the 30-sec hover portion of the task. The pilot cutoff frequency is defined as the half-power frequency of the pilot stick input determined via spectral analysis, and is a measure of the pilot's operating frequency. The cutoff frequency is a good estimate of the pilot-vehicle crossover frequency (Ref. 15).

For the three failure mode cases, the pilot cutoff frequency ω_{co} was higher than the Active configuration indicating that the pilots had to use more high-frequency inputs to control the aircraft position. In both axes, the pilot cutoff frequency was highest for the Light friction configurations and lowest for the High friction configuration. Although this can be an indication of over-controlling in the Light friction configuration, it can also be tied to the fact that of the three failure modes the Light friction configuration had the best performance. In both axes, stick force RMS values were higher for the failure configurations than the Active configuration, and as expected increased from Light to Heavy friction, as more force was required to overcome the stick friction.

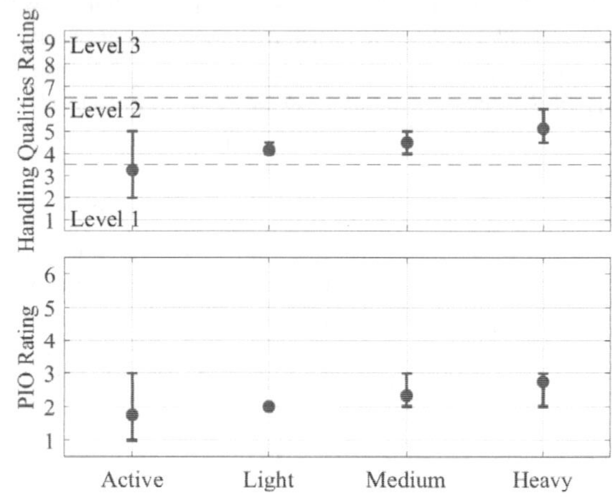

Fig. 12. Handling Qualities and Pilot Induced Oscillation Ratings for the Hover MTE tested in DVE. (Markers indicate average ratings and error bars indicate best and worst ratings.).

Hover MTE (DVE)

Figure 12 shows the average (markers) as well as best and worst (error bars) HQRs and PIORs for the Hover MTE tested in DVE for the active and three failure mode configurations. In DVE, pilots flew this task using control law mode CL2, which provided an ACAH response type in pitch and roll with position hold functionality with stick in detent (or centered in the case of the failure mode configurations), VRCHH in the heave axis, and RCDH in the yaw axis, as shown in Fig. 2.

Figure 13 shows the aircraft position during the 30-sec hover portion of the task, along with the desired and adequate performance bounds. Overall, desired performance was attainable with the Active sidestick configuration, which received a Level 1 average HQR (3.25) and an average PIOR of 1.75.

Desired performance was also achievable with the Light friction configuration, which received an average HQR of 4.2 and an average PIOR of 2.0. Pilots commented that even though they were able to achieve desired performance, there was "still considerable pilot compensation" which drove the Level 2 HQR.

For the Medium friction configuration, one pilot was able to achieve desired performance while two pilots achieved adequate performance, and the configuration was rated an average HQR 4.5 and PIOR 2.3. For the Heavy friction configuration, overall adequate performance was achievable, and the configuration was rated an average HQR 5.1 and average PIOR 2.75. For the Heavy configuration, pilots commented that the stick force was "too heavy," "undesirable," and "made it feel

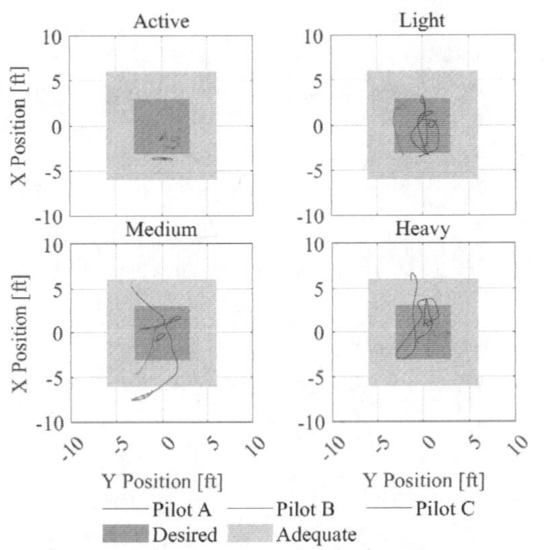

Fig. 13. Position hold performance during 30-sec hover capture portion of the Hover MTE (DVE).

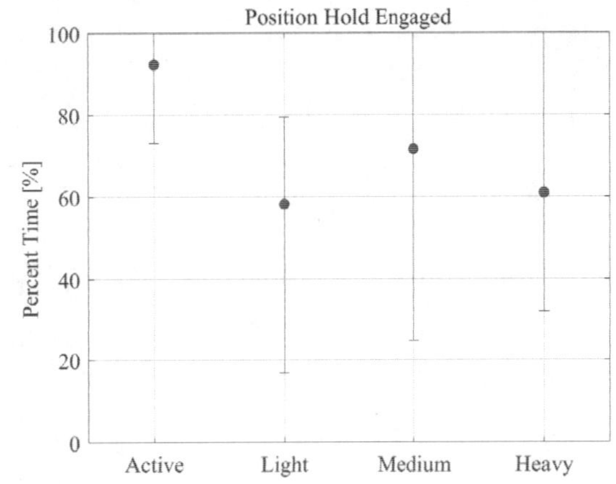

Fig. 14. Percent time of 30-sec hover capture portion of the Hover MTE spent with Position Hold engaged (DVE).

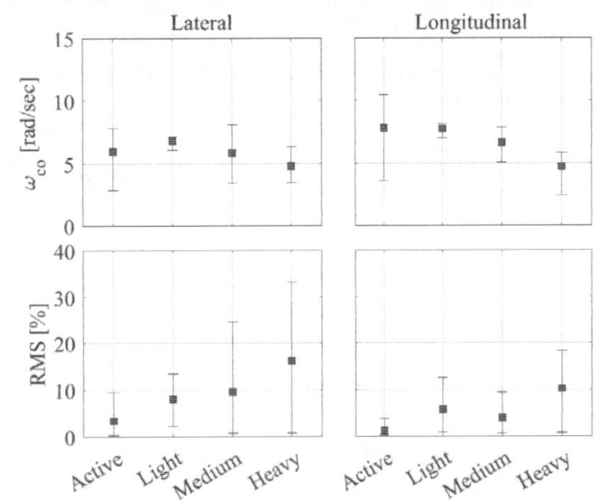

Fig. 15. Pilot force input cutoff frequency and RMS (Hover MTE, DVE).

sluggish." They also commented that the "high friction forces [made] it less harmonious" between the pitch and roll axes.

Figure 14 shows the percent of the 30-sec hover capture portion of the MTE spent with Position Hold engaged. As discussed in the Control Laws section, for Position Hold to engage, the stick must be in the centered, or neutral, position. This is easily accomplished in the Active sidestick configuration due to the detent provided by the gradient and preload forces, however it was not clear ahead of time if pilots would be able to locate the stick neutral position in the failure mode configurations without the gradient and preload forces. Overall, pilots spent the most time with Position Hold engaged in the Active sidestick configuration, however, even though pilots spent less time with Position Hold engaged in the failure mode configurations, they were still on average spending more than 50% of the time with Position Hold engaged. This suggests that pilots were able to find the stick neutral position with the friction only cases and make use of the additional augmentation provided by the control laws. The amount of time spend with Position Hold engaged in the failed state was similar between all three failure mode configurations, suggesting that between the values tested, the amount of friction does not have an appreciable effect on the pilot's ability to find the stick neutral position.

Figure 15 shows the pilot cutoff frequency ω_{co} in the lateral and longitudinal axes as well as the stick RMS in both axes, for pilot stick force F inputs, during the 30-sec hover portion of the task. For the Hover MTE flown in DVE using an attitude command response type with position hold, the pilot cutoff frequencies ω_{co} were similar for all configurations in the lateral axis. In the longitudinal axes, pilot cutoff frequency was highest for the Active and Light friction configurations and lowest for the Heavy friction configuration. As with the Hover MTE flown in GVE using a rate command response type, stick RMS values were lowest for the Active configuration, higher for the Low friction configuration, and highest for the Heavy friction configuration, as more force is required to move the stick.

Pitch Attitude Capture and Hold MTE

Figure 16 shows the average (markers) as well as best and worst (error bars) HQRs and PIORs for the Pitch Attitude Capture and Hold (PACH) MTE tested at 200 kts for the active and three failure mode configurations. Figure 17 shows the PACH task performance and longitudinal stick displacement and force for the different configurations.

Desired performance was achievable for the Active configuration and pilots rated it an average HQR 2.3 and PIOR 1.0. Pilots commented that the Active configuration was "predictable," "consistent," and that they were "not working too hard to get desired [performance]."

Desired performance was also generally achievable for the Light friction configuration and pilots rated it an average HQR

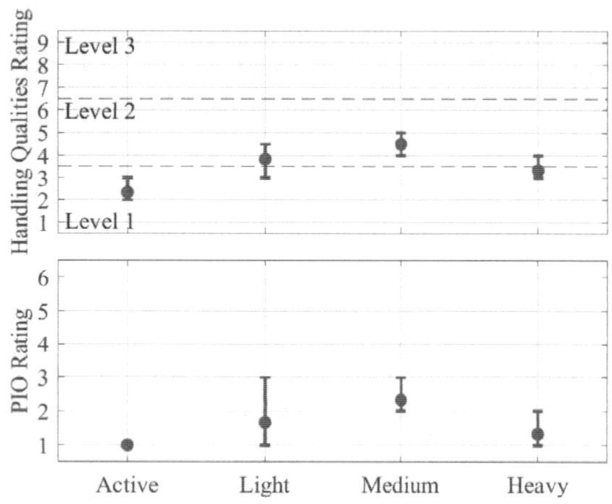

Fig. 16. Handling Qualities and Pilot Induced Oscillation Ratings for the Pitch Attitude Capture and Hold MTE. (Markers indicate average ratings and error bars indicate best and worst ratings.)

3.8 and PIOR 1.7. For the Light friction configuration, pilots commented that there was a "tendency to overshoot" and that even though desired performance was achievable, they were "extremely conscious not to over-control." Furthermore, pilots commented that the "friction seemed too light for the task."

Consistent desired performance was not achievable with the Medium friction configuration and pilots rated it an average HQR 4.5 and PIOR 2.3. Pilots commented that the Medium friction configuration, "felt similar to [the Light friction configuration]," "still too light," and that "both were a little bit challenging." Furthermore, pilots noted having to do a "fair amount of thinking about the lead and controlling the overshoot."

Desired performance was generally achievable for the Heavy friction configuration and pilots rated it an average HQR 3.3 (Level 1) and PIOR 1.3. Although pilots generally preferred the Heavy friction configuration over the Light or Medium friction configurations for this task, they did comment that fine adjustments were less predictable than the other two friction configurations.

Figure 17 also shows the pilot stick displacement δ_{lon} and force F_{lon} inputs for the Active and three failure mode configurations. For the Active sidestick configurations, pilots were able to return the stick to detent and relax the force on the stick during the steady attitude hold portion of the MTE. For the three failure mode configurations, stick displacement had more excursions from neutral during the steady attitude hold portion of the MTE. This can also be seen in the stick force input, which has more high-frequency excursions from zero force for the three failure mode configurations. However, even with the more frequent pilot inputs, no PIOs were encountered for any of the failure mode configurations tested.

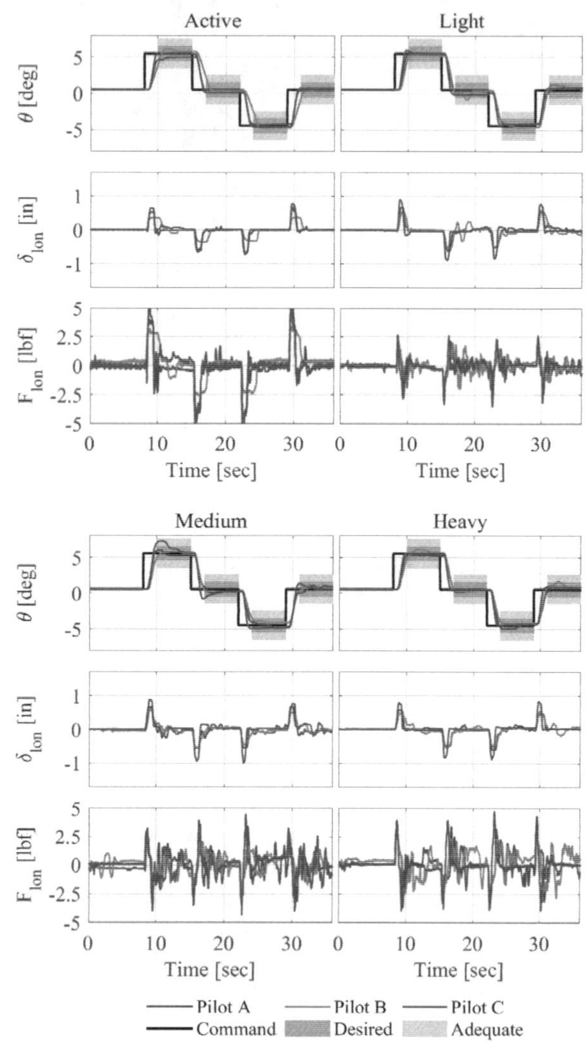

Fig. 17. Pitch Attitude Capture and Hold MTE task performance and pilot stick input.

Bank Angle Capture and Hold MTE

Figure 18 shows the average (markers) as well as best and worst (error bars) HQRs and PIORs for the Bank Angle Capture and Hold (BACH) MTE tested at 200 kts for the active and three failure mode configurations. Figure 19 shows the BACH task performance and lateral stick displacement and force for the different configurations.

Desired performance was achievable for the Active configuration and pilots rated it an average HQR 3.0 and PIOR 1.3. Similar to the PACH MTE, pilots commented that the Active configuration was "predictable," "consistent," and that there were "no issues with the mechanical characteristics."

In general, desired performance was also achievable for the three failure mode configurations, although each configurations did also have excursions into adequate performance. The

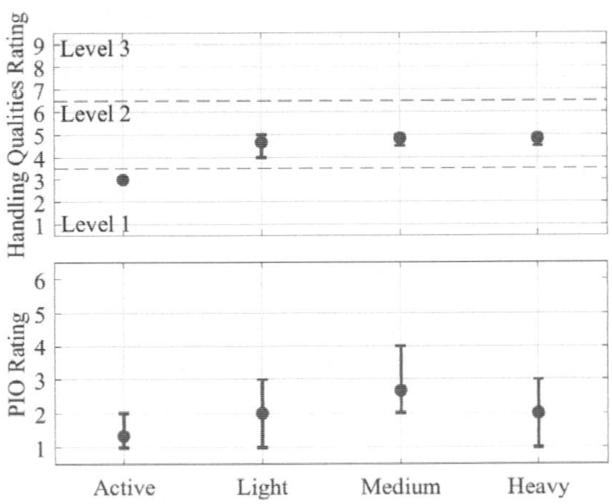

Fig. 18. Handling Qualities and Pilot Induced Oscillation Ratings for the Bank Angle Capture and Hold MTE. (Markers indicate average ratings and error bars indicate best and worst ratings.)

Light friction configuration was rated an average HQR 4.7 and PIOR 2.0. Pilots commented that the Light friction configuration had "more tolerable forces" than the Medium or Heavy friction configurations, and that they "did not feel any tendency to cause oscillations."

The Medium friction configuration was rated an average HQR 4.8 and PIOR 2.7. Pilots commented that they "[were] working hard" and "could...get tired flying" this configuration. The Heavy friction configuration was rated an average HQR 4.8 and PIOR 2.0. Pilots commented that they "felt like friction was too high" with the Heavy friction configuration, that "it [was] easy to get off axis inputs," and that they had to "tone down" their aggressiveness.

Figure 19 also shows the pilot stick displacement δ_{lat} and force F_{lat} inputs for the Active and three failure mode configurations. As with the Pitch Attitude Capture and Hold MTE, for the Active sidestick configurations, pilots were able to return the stick to detent and relax the force on the stick during the steady attitude hold portion of the MTE. For the three failure mode configurations, stick displacement and force had more excursions from neutral during the steady attitude hold portion of the MTE, however no PIOs were encountered for any of the configurations tested.

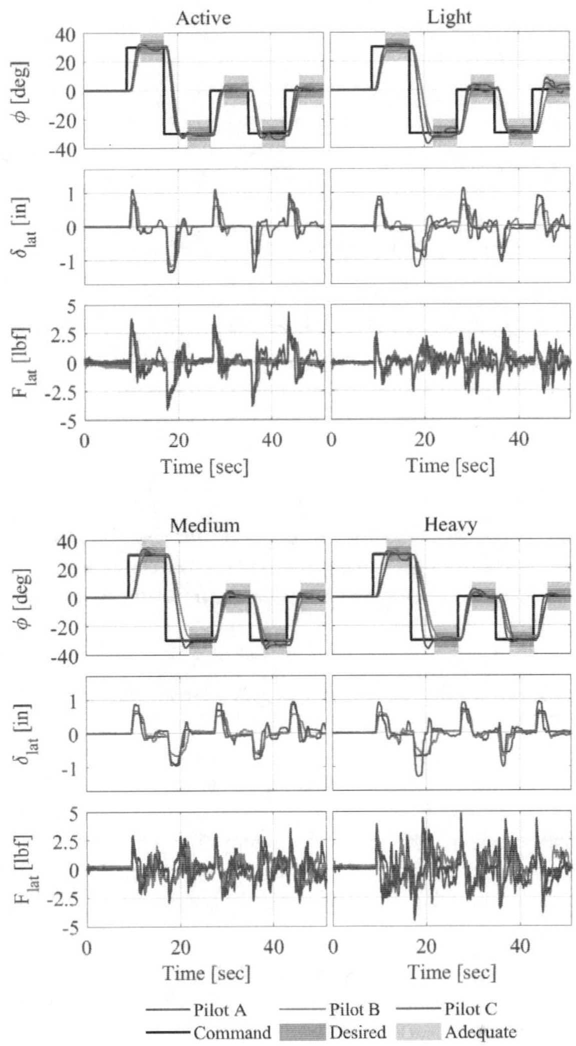

Fig. 19. Bank Angle Capture and Hold MTE task performance and pilot stick input.

Station Keeping MTE

Figure 20 shows the average (markers) as well as best and worst (error bars) HQRs and PIORs for the Station Keeping MTE tested at 200 kts for the active and three failure mode configurations. Pilots flew this MTE with Velocity Hold mode engaged, thus making it primarily a lateral and longitudinal axis task, for control left/right and up/down position behind the tanker, respectively. Figure 21 shows the relative left/right and up/down position between the ownship and tanker aircraft during the 30-sec station keeping portion of the MTE.

Desired performance was achievable for the Active configuration and pilots rated it an average HQR 3.3 and PIOR 1.7. Pilots commented that there was some compensation involved in "trying to back out of the loop and null out any perturbations."

Desired performance was also achievable for the three failure mode configurations, but all three also had runs with excursions into adequate performance. The Light friction configuration was rated an average HQR 4.3 and PIOR 2.7. Pilots commented that with the Light friction configuration they were able to achieve desired performance but with "more than moderate pilot compensation." Pilots also commented that the response "was pretty sensitive in pitch" and that the friction was "too light."

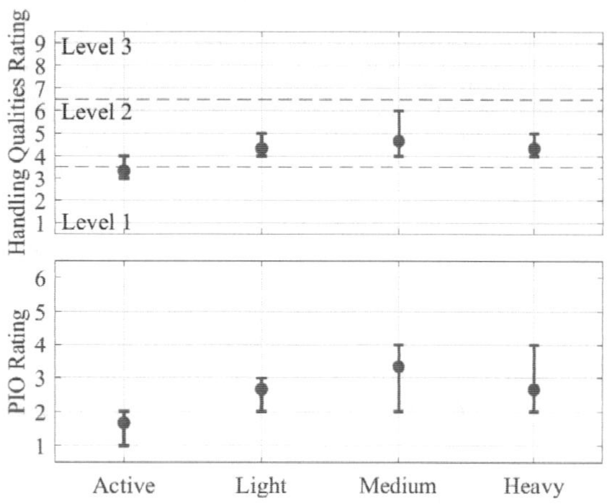

Fig. 20. Handling Qualities and Pilot Induced Oscillation Ratings for the Station Keeping MTE. (Markers indicate average ratings and error bars indicate best and worst ratings.)

The Medium friction configuration was rated an average HQR 4.7 and PIOR 3.3, and pilots commented that they were "more fatigued than the [Light friction] configuration." The Heavy friction configuration was rated an average HQR 4.3 and PIOR 2.7. Pilots commented that the "weight of the friction probably prevented [them] from getting into more of a PIO," that the "forces are heavy, but [they] could stabilize a little bit easier," and that they "[liked] this better than the Medium [friction configuration]."

Figure 22 shows the aircraft roll and pitch rate RMS during the 30-second station keeping portion of the MTE. The three failure mode configurations had similar aircraft motion (RMS values), and all were higher than the Active sidestick configurations.

Figure 23 shows the pilot cutoff frequency ω_{co} in the lateral and longitudinal axes as well as the stick RMS in both axes, for pilot stick force F inputs. The three failure mode configurations had generally similar values of pilot cutoff frequency ω_{co}, and all are higher than the Active configuration indicating that the pilots had to use more high-frequency inputs to control the relative aircraft position with the failure mode configurations. The stick force RMS values are relatively lower for the Active and Light friction configurations, and highest for the Heavy friction configuration.

Operational Scenario

Figures 24 through 26 show example time histories of the operational scenario for the Light and Heavy friction failure mode configurations. Figure 24 shows the aircraft track through the defined corridor, with markers placed where the sidestick failure was injected. In both cases, pilots were able to continue flying the scenario, remain within the corridor, and

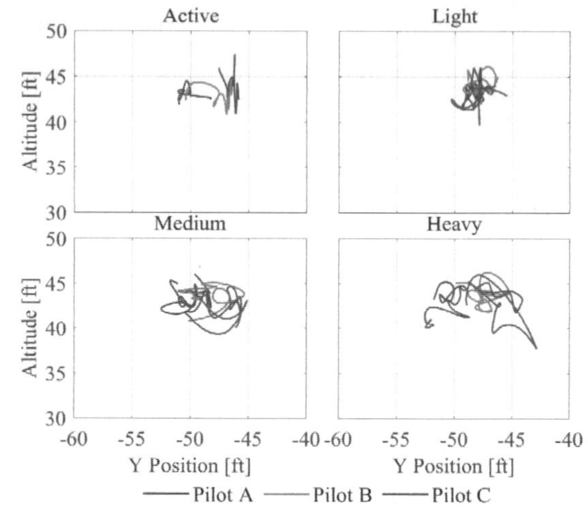

Fig. 21. Relative ownship position during 30-sec station keeping portion of the Station Keeping MTE.

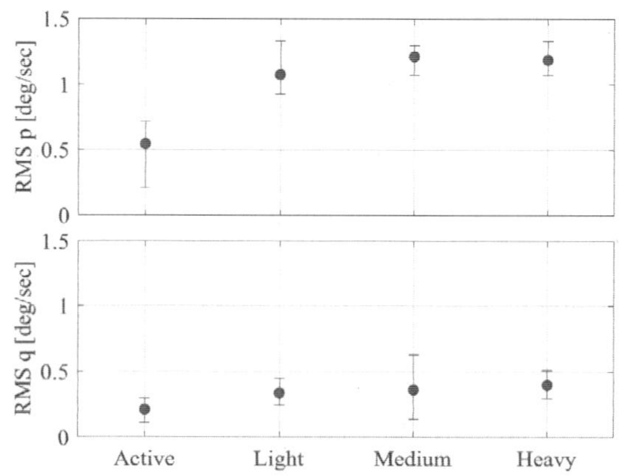

Fig. 22. Aircraft roll and pitch rate RMS during 30-sec station keeping portion of the Station Keeping MTE.

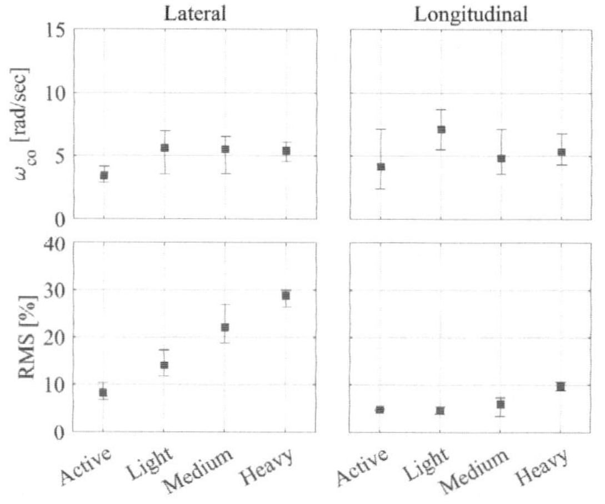

Fig. 23. Pilot force input cutoff frequency and RMS (Station Keeping MTE).

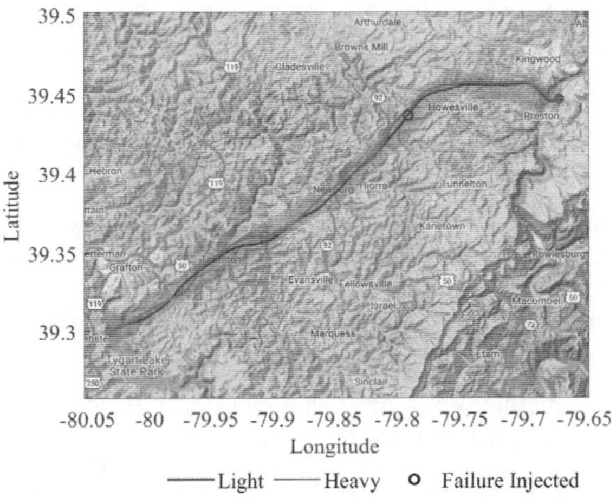

Fig. 24. Operational scenario results, map view.

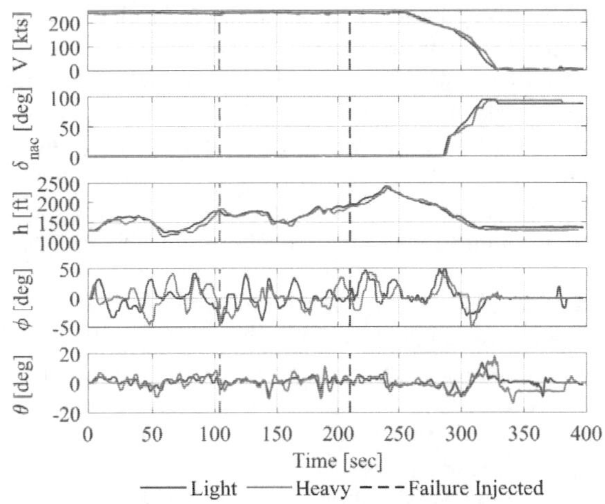

Fig. 25. Operational scenario results, aircraft states.

come to a stable hover at the defined point at the end of the corridor.

Figure 25 shows aircraft airspeed V, nacelle tilt angle δ_{nac}, altitude h, bank angle ϕ, and pitch attitude θ time histories, and Fig. 26 shows the lateral and longitudinal sidestick force and displacement inputs.

As with the MTEs, more high-frequency force input can be seen in both axes after the sidestick failure was injected. In addition, as expected, larger overall force inputs were required with the Heavy friction configuration compared to the Light friction configuration. However, in neither case was a PIO encountered, and no additional oscillations were seen in the aircraft attitudes after the sidestick failures were injected.

Since the operational scenario did not have defined desired or adequate performance bounds, HQRs and PIORs were not collected. Instead, only pilot comments were collected. For all failure cases, pilots were able to continue flying the scenario and convert to a hover at the end, and the failure transients were not objectionable. For the Light friction configuration, pilots commented that the stick felt "floppy" and they had to "tighten [their] grip" and "stay in the loop more." Pilots also commented that "making flight path changes [was] less predictable," although there was no sense of over-controlling.

For the Medium friction configuration, pilots commented that it was "less fatiguing" as compared to the Heavy friction configuration and felt "comfortable." Finally, for the Heavy friction configuration, pilots commented that it was "a lot of work" to fly and they had "to be attentive," although it was not objectionable.

Overall, pilots were able to continue flying the operational scenario with aggressive maneuvering and convert to hover with all three failure mode configurations. Based on the pilot comments, the Medium friction configuration was a good compromise between less predictable aircraft response (Light friction) and more fatiguing (Heavy friction).

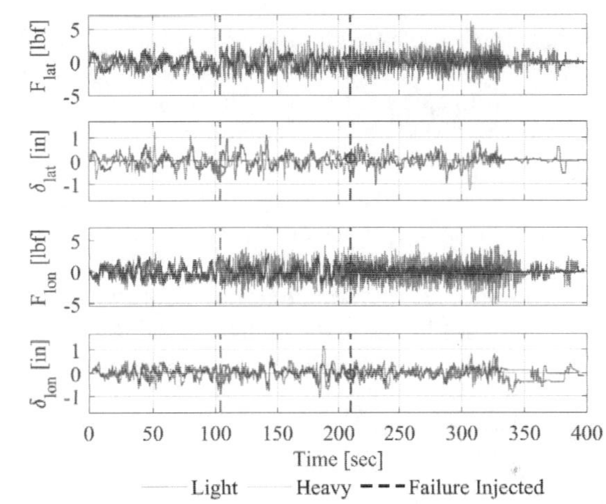

Fig. 26. Operational scenario results, pilot inputs.

CONCLUSIONS

A piloted simulation study was conducted in the NASA Ames Vertical Motion Simulator to assess handling qualities of an active sidestick failure mode resulting in a friction only force feel. Five Mission Task Elements (MTEs) and one operational scenario were used to evaluate the Active sidestick configuration, as well as three failure mode configurations with varying levels of friction (Light, Medium, and Heavy). The results support the following conclusions:

1. Overall Level 1 handling qualities were achievable with the Active sidestick configuration. This validated the control law architecture and optimization strategy used, and set a known Level 1 baseline for the failure mode evaluations.
2. Level 2 handling qualities were achievable for both hover and high-speed MTEs during sidestick failure with the

friction values evaluated (0.4, 0.9, and 1.4 lbf). Additionally, no Pilot Induced Oscillations were encountered with the friction values evaluated. Based on these results, the minimum control force gradients required for Level 2 for sidesticks listed in MIL-DTL-32742 could be relaxed from 0.5 lbf/in to 0 lbf/in with an acceptable amount of friction.

3. The Light friction configuration (0.4 lbf) was preferred over Medium (0.9 lbf) and Heavy (1.4 lbf) friction configurations during evaluations of the Hover MTE in both Good and Degraded Visual Environments.
4. When evaluating the Hover MTE in DVE with the sidestick failure mode configurations, pilots were able to find the stick neutral position even without a restoring gradient or detent, enabling use of the position hold function. Therefore, hold modes that require the stick to be centered should be enabled during a friction only sidestick failure mode. Although it was not evaluated here, an additional stick position display could be helpful in finding the stick neutral position during a loss of force feel failure.
5. At high speed, pilots preferred the Heavy friction configurations (1.4 lbf) for the Pitch Attitude Capture and Hold and Station Keeping MTEs. For the Bank Angle Capture and Hold MTE, pilots preferred the Light friction configurations (0.4 lbf).
6. After sidestick failure during the operational scenario, pilots were able to continue high-speed terrain flight and accomplish a convert to hover with all three failure mode configurations. During operational scenario evaluations, pilots preferred the Medium friction configuration (0.9 lbf), which was a good compromise between less predictable aircraft response (Light friction) and more fatiguing (Heavy friction).

ACKNOWLEDGMENTS

The authors would like to sincerely thank the entire NASA Ames VMS team, especially Emily Lewis, Matthew Blanken, and Peter Zaal. In addition, the authors would like to thank the pilots who participated in this study for providing their time, insight, and excellent comments, as well as the flight test engineers who contributed to this effort.

REFERENCES

[1]Berger, T., , Dotson, B. N., Gong, A., Lusardi, J. A., Mansur, M. H., Ott, C. R., Ogden, W. R., and Miller, M. L., "Effects of Linked vs. Unlinked Cyclic Controllers on Crew Coordination," Vertical Flight Society 79th Annual Forum Proceedings, West Palm Beach, FL, May 2023.

[2]Anon., "Aerospace Active Inceptor Systems for Aircraft Flight and Engine Controls, Aerospace Recommended Practive For," SAE-ARP5764, February 2013.

[3]Anon., "Handling Qualities for Military Rotorcraft," MIL-DTL-32742, Department of Defense Detail Specification, March 2023.

[4]Einthoven, P., "Active Controller Performance Requirements," Vertical Flight Society 60th Annual Forum Proceedings, Baltimore, MD, May 2004.

[5]Müllhäuser, M. and Barnett, M., "Development Of A Generic Test For Transient Recovery Handling From Helicopter Active Inceptor System Failures To A Near-Zero Control Force Condition," Vertical Flight Society 71st Annual Forum Proceedings, Virginia Beach, VA, May 2015.

[6]Anon., "Military Specification: Flying Qualities of Piloted V/STOL Aircraft," MIL-F-83300, December 1970.

[7]Juhasz, O., Celi, R., Ivler, C. M., Tischler, M. B., and Berger, T., "Flight Dynamic Simulation Modeling of Large Flexible Tiltrotor Aircraft," Vertical Flight Society 68th Annual Forum Proceedings, Fort Worth, TX, May 2012.

[8]Celi, R., "HeliUM 2 Flight Dynamic Simulation Model: Developments, Technical Concepts, and Applications," Vertical Flight Society 71st Annual Forum, Virginia Beach, VA, May 2015.

[9]Berger, T., Juhasz, O., Lopez, M. J. S., Tischler, M. B., and Horn, J. F., "Modeling and Control of Lift Offset Coaxial and Tiltrotor Rotorcraft," *CEAS Aeronautical Journal*, Vol. 11, (1), January 2020, pp. 191–215.

[10]Tobias, E. L. and Tischler, M. B., "A Model Stitching Architecture for Continuous Full Flight-Envelope Simulation of Fixed-Wing Aircraft and Rotorcraft from Discrete-Point Linear Models," U.S. Army AMRDEC Special Report RDMR-AF-16-01, April 2016.

[11]Berger, T., Blanken, C. L., Lusardi, J. A., Tischler, M. B., and Horn, J. F., "Tiltrotor Flight Control Design and High-Speed Handling Qualities Assessment," *Journal of the American Helicopter Society*, Vol. 67, (3), pp. 114-128(15), DOI: 10.4050/JAHS.67.032009, July 2022.

[12]Tischler, M. B., Berger, T., Ivler, C. M., Mansur, M. H., Cheung, K. K., and Soong, J. Y., *Practical Methods for Aircraft and Rotorcraft Flight Control Design: An Optimization-Based Approach*, AIAA, Reston, VA, 2017.

[13]Danek, G. L., "Vertical Motion Simulator Familiarization Guide," Tech. rep., NASA-TM-103923, 1993.

[14]Klyde, D. H. *et al.*, "Piloted Simulation Evaluation of Attitude Capture and Hold MTEs for the Assessment of High-Speed Handling Qualities," Vertical Flight Society 74th Annual Forum Proceedings, Phoenix, AZ, May 2018.

[15]Tischler, M. B. and Remple, R. K., *Aircraft and Rotorcraft System Identification: Engineering Methods and Flight Test Examples Second Edition*, AIAA, 2012.

NRC·CNRC

An Analytical Study of Helicopter Handling Qualities Characteristics due to Active Force-Sensing Side-Stick Utilization

NRC Flight Research Laboratory (Ottawa, Canada)

Publication: VFS.HQ.TM 2025

Marc Alexander, Senior Flight Research Officer

January 2025

National Research Council Canada Conseil national de recherches Canada

Overview

- ❑ Rationale, Objectives, & Methodology
- ❑ Test Vehicle Systems Descriptions
- ❑ Side-Stick Engineering (Integration, Identification, Research Data Set)
- ❑ Study Results on Side-Stick Tuning Processes:
  - Hover Stabilization Analysis
  - Attitude Capture Analysis
  - Mission Task Analysis
  - Comparative Analysis of Parameters
- ❑ Closing Remarks

Rationale: Next Generation Capabilities & Aircraft Interface Requirements

Capability Enhancements afforded by automation, autonomous, & intelligent flight systems:

- ❑ Operations: Mission Task Execution
- ❑ Availability: Environmental Robustness
- ❑ Life Extension: Condition Avoidance

Resulting requirements for advanced aircrew interface technology:

- ❑ Assistance: Cueing and Augmentation

Objectives: Assessment of Side-Stick Tuning Processes for Handling Qualities Attainment

Application NRC variable stability helicopter flight research facilities:

- ❑ To assess side-stick tuning processes for attaining desired flight characteristics for mission task execution
- ❑ To identify inter-relationship between stick parameters and mission task performance

Systems Descriptions: Test Aircraft and Aircrew Interface Systems

Test Aircraft: NRC Bell 412 Advanced Systems Research Aircraft

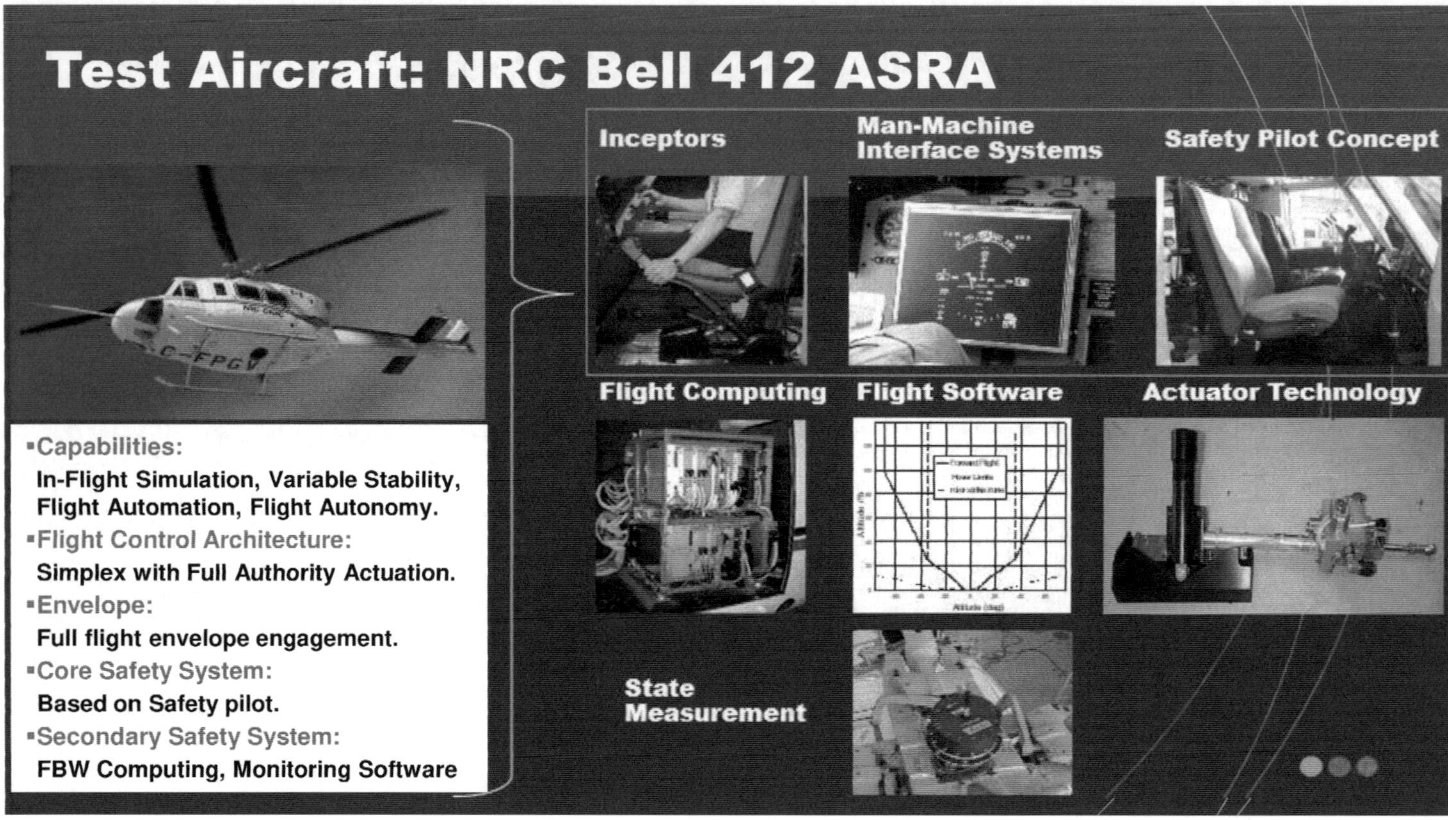
Test Aircraft: NRC Bell 412 ASRA
Inceptors
Man-Machine Interface Systems
Safety Pilot Concept
Flight Computing
Flight Software
Actuator Technology
State Measurement
C-FPGV
▪Capabilities:
In-Flight Simulation, Variable Stability, Flight Automation, Flight Autonomy.
▪Flight Control Architecture:
Simplex with Full Authority Actuation.
▪Envelope:
Full flight envelope engagement.
▪Core Safety System:
Based on Safety pilot.
▪Secondary Safety System:
FBW Computing, Monitoring Software

NRC Bell412 ASRA Aircrew Interface Systems:
Active Side-Stick (Stirling Dynamics Inc.)

Systems Implementation: Integration, Identification, Characterization

Side-Stick Integration:

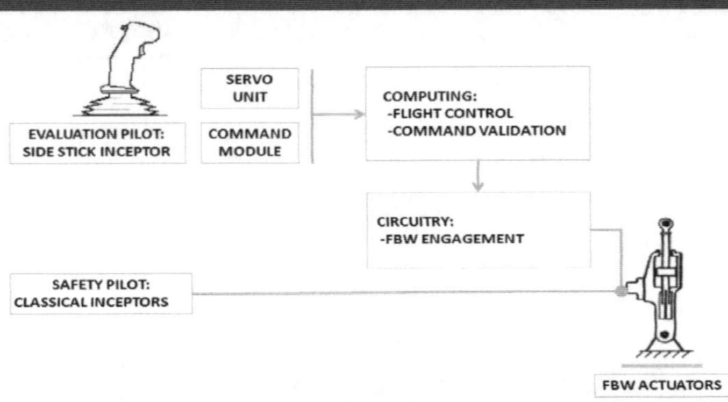

Side-Stick (SS) Basic Design Elements:	
SS Maximum Deflection Range (Rf: x,y)	±25±0.2 Deg.
SS Maximum Force (Rf: Fx, Fy @ FRP)	150 N.
SS Deflection Stop Asymmetry (Rf: y>x)	30%
SS Force-Gradient Asymmetry (Rf: x>y)	70%
Notes: FRP = Finger Reference Point ; Rf = Reference;	

Side-Stick Identification:

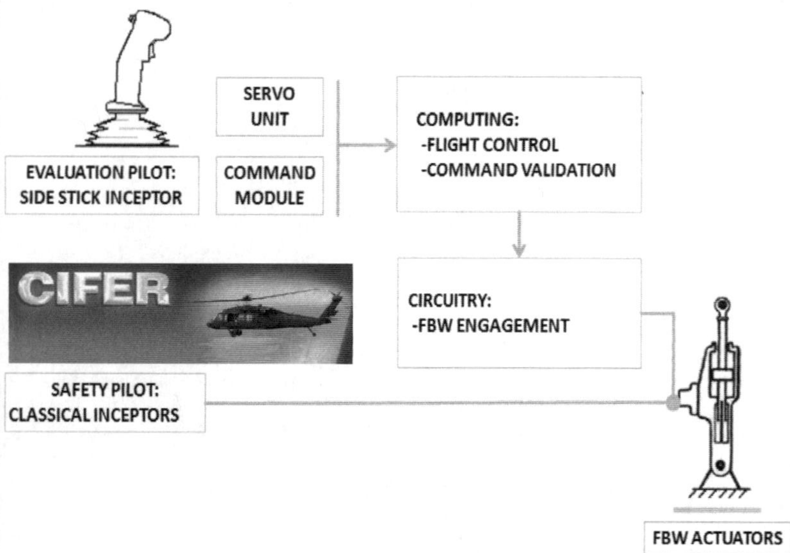

Side-Stick Characterization:

Cueing	GVE[a]
Response Type	Rate Damped, Axes (Roll, Pitch)[b,c]

[a]Good Visual Environment (GVE)
[b]Yaw Response: Rate Damped (RD)
[c]Heave Response: Open-Loop, Damped, Height-Hold (HTH)

Baseline Side-Stick Setting Cases:	δ, mm (Deg)		k_{FG}, N/mm, (lb/in)	
	x	y	x	y
Low Gradient/Moderate Throw: Case, A	18.71 (7.05)	24.94 (9.4)	0.52 (2.98)	0.4 (2.2)
Low Gradient/High Throw: Case, B	35.81 (13.5)	47.75 (18)	0.21 (1.2)	0.16 (0.9)
Moderate Gradient and Throw: Case, D	17.9 (6.75)	23.87 (9)	1.46 (8.37)	1.1 (6.3)

Side-Stick Force Basis: ○

Side-Stick Displacement Basis: ◇

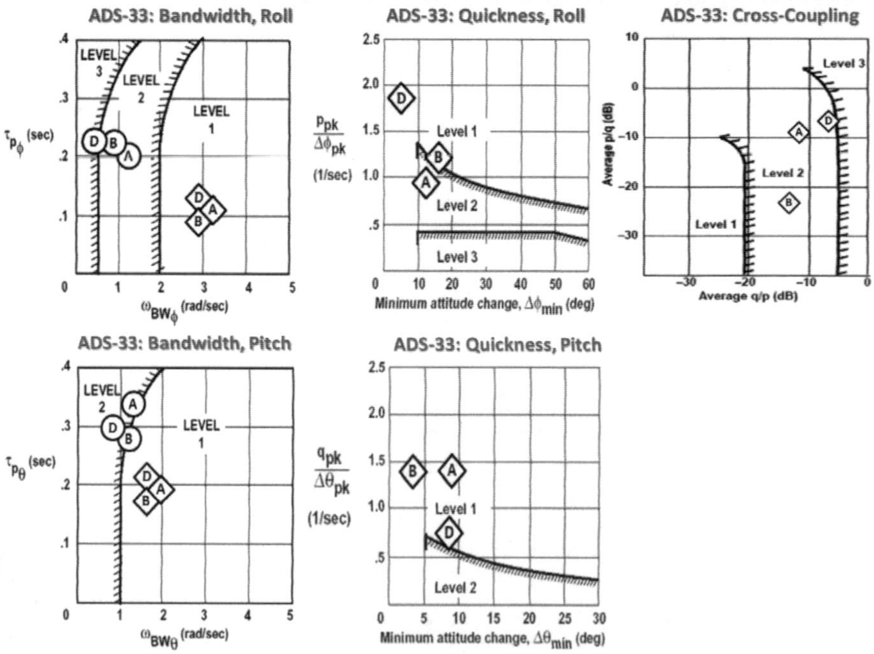

Research Methodology: Flight Test Data Set

Research Methodology & Data Set

Location: NRC ADS-33 Low Speed Test Area

Tasks: Functional integration testing, Test pilot school exercises, and Mission task evaluations.

Content: Over 100 events performed by 9 pilots.

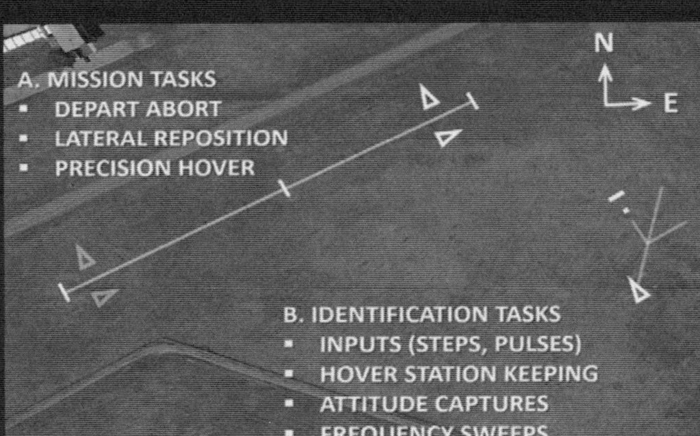

Side-Stick Integration, Identification, Characterization (Open-Loop, Closed-Loop)

Data Set Selection
- Research Focus: Side-Stick Tuning Processes for Handling Qualities Mission Task Attainment

Data Studies
- Hover Stabilization Analysis (Parameter Bounds)
- Attitude Capture Analysis (Workload Assessment; Control Activity, Strategy)
- Mission Task Analysis (Tuning Progression)
- Comparative Analysis (Preferences)

Research Results: Side-Stick Tuning and Optimization Processes

Hover Stabilization Study

Test Cases:
- MTE's: 16 mission tasks (Precision Hover, Lateral Reposition, Depart Abort)

Analysis:
- Assessment of side-stick design parameters suitable for achieving minimal aircraft positional errors during MTE stabilization phases.

Task performance:
- Time-to-stable ($t \leq 25$ seconds)
- Position errors (longitudinal, lateral error, $e \leq 3$ft)

Metrics:
- SS Parameter Variance & MTE Success Rate

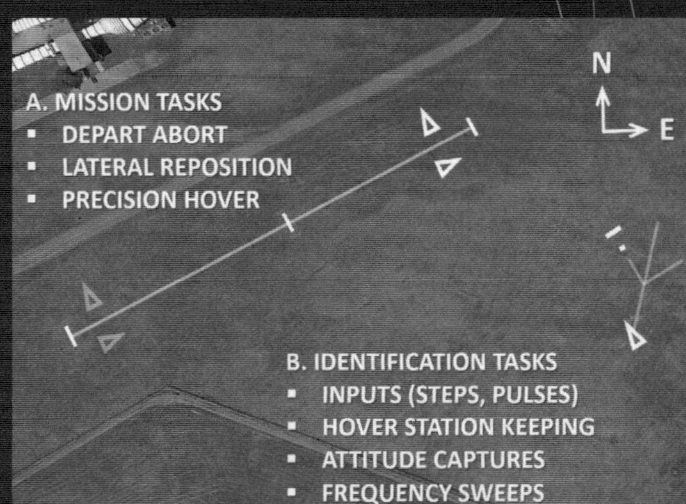

Hover Stabilization Study, Lateral Axis:

Side-Stick Configuration Parameters	Parameter Value Bounds & Error Tolerance Success Rate	
	x-axis	y-axis
F_{max}	[8.2 – 9.9] , 31%	[4.1 – 9.5] , 38%
δ_{max}	[0.9 – 4.2] , 31%	[1.1 – 7.9] , 44%
k_{FGmax}	[0.8 – 8.5] , 38%	[0.6 – 4.2] , 63%
$S_{\delta max}$	[0.3 – 4.2] , 38%	[0.2 – 3.9] , 63%
S_{Fmax}	[1.3 – 3.7] , 38%	[0.4 – 3.0] , 63%

Notes:
Format ([l – m], n%) = ([Parameter Range], n of 16 in e ≤ 3ft)
F = N; δ = mm; k_{FG} = N/mm; S_δ = mm/(Deg/s); S_F= N/(Deg/s)

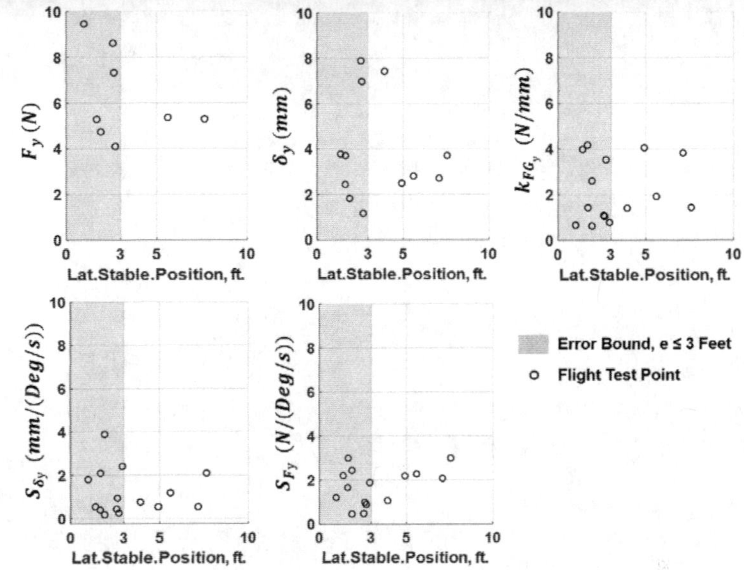

Hover Stabilization Study, Longitudinal Axis:

Side-Stick Configuration Parameters	Parameter Value Bounds & Error Tolerance Success Rate	
	x-axis	y-axis
F_{max}	[8.2 – 9.9] , 31%	[4.1 – 9.5] , 38%
δ_{max}	[0.9 – 4.2] , 31%	[1.1 – 7.9] , 44%
k_{FGmax}	[0.8 – 8.5] , 38%	[0.6 – 4.2] , 63%
$S_{\delta max}$	[0.3 – 4.2] , 38%	[0.2 – 3.9] , 63%
S_{Fmax}	[1.3 – 3.7] , 38%	[0.4 – 3.0] , 63%

Notes:
Format ([l – m], n%) = ([Parameter Range], n of 16 in e ≤ 3ft)
F = N; δ = mm; k_{FG} = N/mm; S_δ = mm/(Deg/s); S_F= N/(Deg/s)

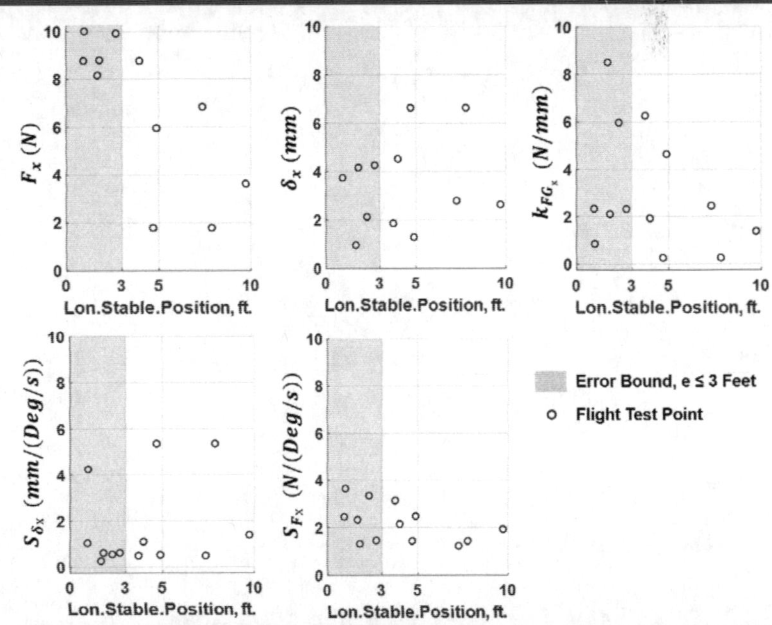

Hover Stabilization Study

Key Observations :

- Inherent aircraft coupling
- Asymmetric human interface force-displacement (ergonomic, anthropomorphic) side-stick configuration requirements
- Variability illustrated the interrelationship between stick parameters and mission task performance.

Side-Stick Configuration Parameters	Parameter Value Bounds & Error Tolerance Success Rate	
	x-axis	y-axis
F_{max}	[8.2 – 9.9] , 31%	[4.1 – 9.5] , 38%
δ_{max}	[0.9 – 4.2] , 31%	[1.1 – 7.9] , 44%
k_{FGmax}	[0.8 – 8.5] , 38%	[0.6 – 4.2] , 63%
$S_{\delta max}$	[0.3 – 4.2] , 38%	[0.2 – 3.9] , 63%
S_{Fmax}	[1.3 – 3.7] , 38%	[0.4 – 3.0] , 63%

Notes:
Format ([l – m], n%) = ([Parameter Range], n of 16 within e ≤ 3ft)
F = N; δ = mm; k_{FG} = N/mm; S_δ = mm/(Deg/s); S_F= N/(Deg/s)

Low Speed Attitude Capture Study

Test Cases:
- Attitude Capture Sequence

Analysis:
- Assessment of impact of sidestick design parameters on handling qualities

Task performance:
- Low Stick Throw: 7% generating p = 13 deg/s.
- Hih Stick Throw: 45% generating p = 15 deg/s.

Metrics:
- SS Parameters & workload metrics (control strategy, task performance)

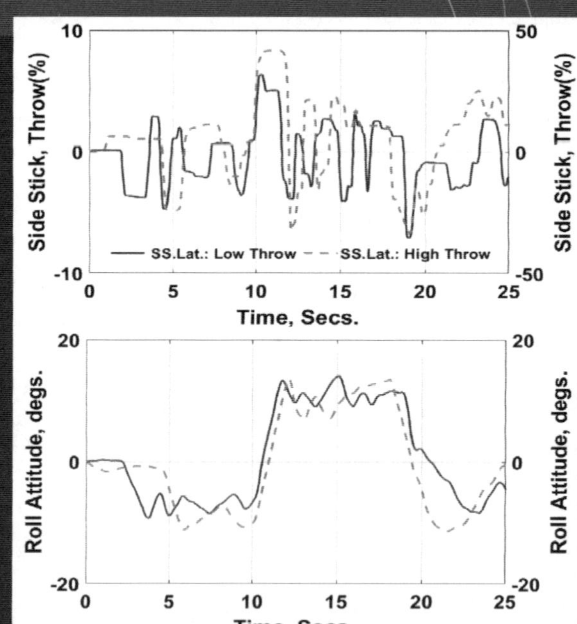

Low Speed Attitude Capture Study, Stability & Workload

Key Observations :

- Changes in workload (compensation, activity, strategy) with stick displacement-force limits.
- Areas included:
  - Magnitude of Axis Cross-Coupling with throw
  - Magnitude of Control Activity (Reversals, Displacement Rate) with throw;
    - Low throw, higher input displacement rate (5 to 12 deg/s^2)
    - High throw, lower input displacement rate (2 to 5 deg/s^2)

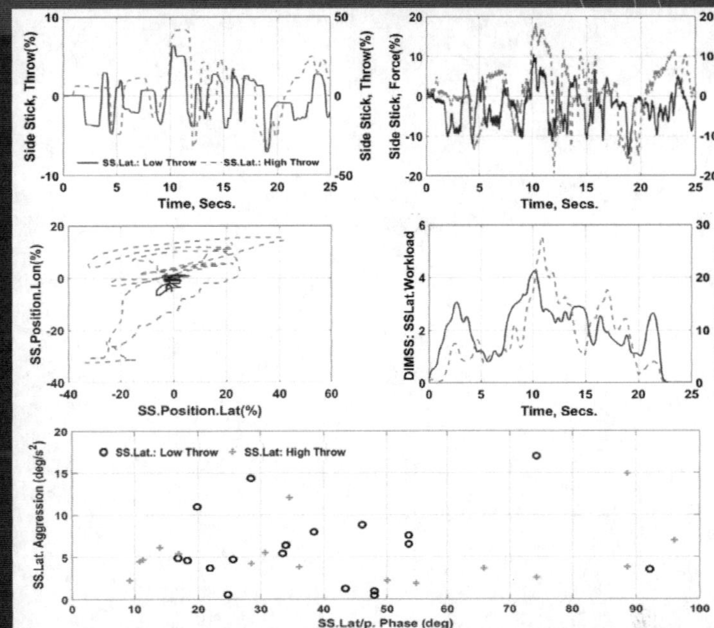

Low Speed Attitude Capture Study, Pilot Control Strategy

Key Observations :

- Changes in pilot input control strategy :
  - Low frequency inputs (0.05 to 0.3 Hz) impacting vehicle trim motion.
  - Low throw, higher rate input forces (2x10^7 to 6x10^7 dB)
  - High throw, lower rate input forces (5x10^7 to 12x10^7 dB)

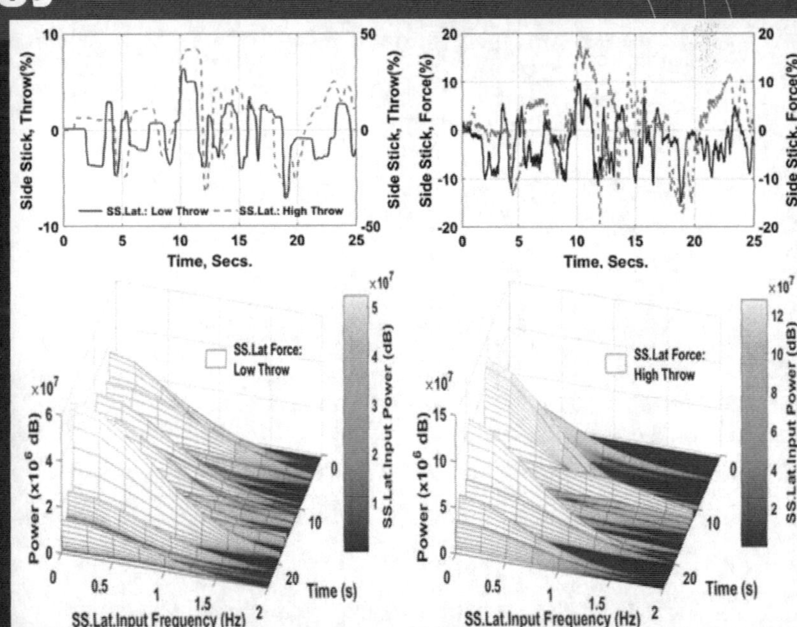

Mission Task Study

Test Cases:
- MTE's: Full Mission Task Analysis (112 events) and Subset Mission Task Analysis (16 events)

Analysis:
- Assessment of side-stick configuration tuning progression and trends.

Task performance:
- Mixed (Identification Tasks, Mission Tasks)

Metrics:
- SS Parameter Variance

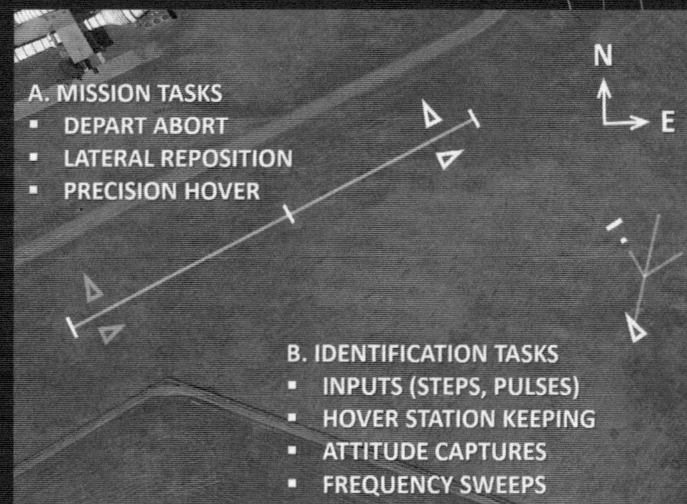

Mission Task Study: N_{MTE} = 112 to 16

- Tuning progression:
  - For Full MTE Analysis (N_{MTE} = 112), tuning progression for gearing, damping, force gradient, and throw [STDlon, STDlat] = [1.8 to 2.8]. STD for sidestick gradient to throw ratio was 1.1.
  - For Subset MTE Analysis (N_{MTE} = 16), tuning progression for gearing, damping, force gradient, and throw [STDlon, STDlat] = [0.1 to 0.79]. STD for sidestick gradient to throw ratio was 0.6.

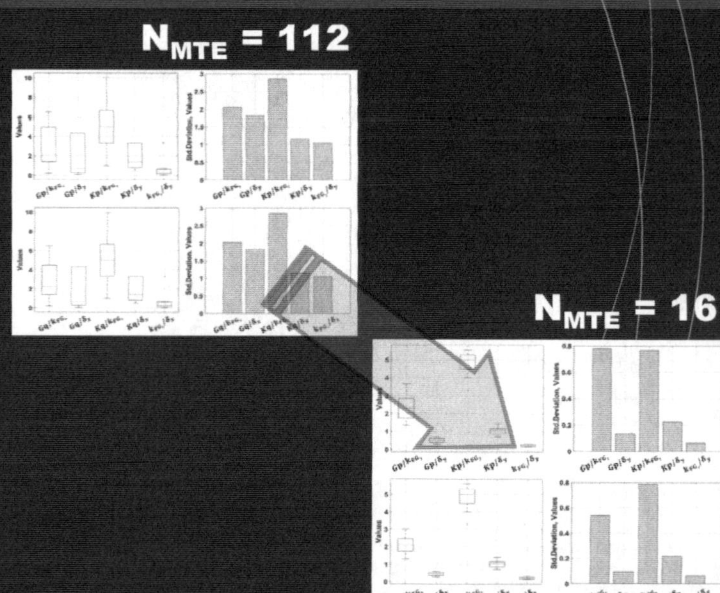

Mission Task Study

Key Observations :

- Tuning processes observed included:
  - Influence of task cut-off frequency relative to task and control bandwidth as key stick configuration factors.
  - Limited influence of force-displacement parameters on bandwidth modification
  - Significant influence of force-displacement parameters on attitude quickness and cross-coupling phenomena

N_{MTE} = 112

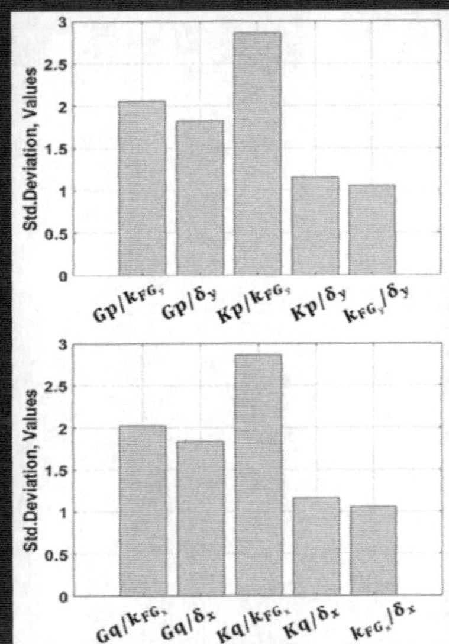

N_{MTE} = 16

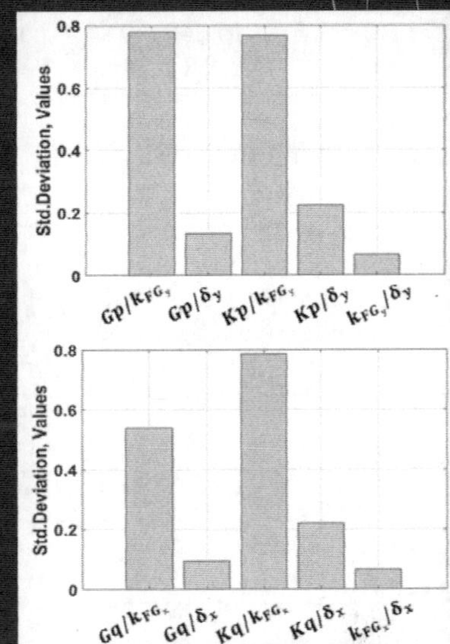

Side-Stick Configuration Comparative Analysis: NRC to Published DLR Research Data

NRC Bell 412 ASRA (Sidestick Fight Test Data):

- MTOW = 5397 kg
- Soft Inplane Hingeless, (b = 4, dMR = 48ft)
- Cargo-Utility
- Hinge Offset (f, l) = (4.6, 9.3) %
- Bandwidth (q, p) = (1.8, 3) rad/s
- Research: Control Power (q, p) = (15, 25) deg/s

DLR EC135 FHS (Sidestick HELIFLY Simulator Test Data):

- MTOW = 5798 kg
- Bearingless (b = 4, dMR = 33.5ft)
- Cargo-Utility
- Hinge Offset, Equivalent = 8.7 %
- Bandwidth (q, p) = (1.8, 3.7) rad/s
- Research: Control Power (q, p) = (31.7, 53) deg/s

Side-Stick Configuration Comparative Analysis: Parameter Preferences

Static Parameters:

x-axis	F_{Bx}	δ_x	k_{FGx}	$S_{\delta x}$	S_{Fx}
NRC	5	19.34	0.43	1.29	0.53
DLR	1.6	41.21	0.4	1.3	0.5
y-axis	F_{By}	δ_y	k_{FGy}	$S_{\delta y}$	S_{Fy}
NRC	3	25.79	0.32	1.03	0.32
DLR	1.1	42.4	0.27	0.8	0.2

Notes:
F = N; δ = mm; k_{FG} = N/mm; S_δ = mm/(Deg/s); S_F = N/(Deg/s)

Dynamics Parameters:

x-axis	ζ_x	ω_{Nx}	I_x	$\tau_{eff.x}$
NRC	0.65	15.50	1.81	0.08
DLR	0.9	23.9	0.7	0.08
y-axis	ζ_y	ω_{Ny}	I_y	$\tau_{eff.y}$
NRC	0.65	14.00	1.71	0.09
DLR	0.9	23.9	0.5	0.08

Notes:
ω_N = rads/s; I = kg; τ_{eff} = secs.

Side-Stick Configuration Comparative Analysis: Projected Handling Qualities

Handling Qualities Projection (DLR Criteria):

DLR FHS EC135:
- Level 1: Sensitivities (Force, Displacement)
- Level 1: Breakout Forces
- Level 1: Damping & Natural Frequency

NRC Bell 412 ASRA:
- Level 1: Sensitivities (Force, Displacement)
- Level 1/2: Breakout Forces
- Level 2: Damping & Natural Frequency

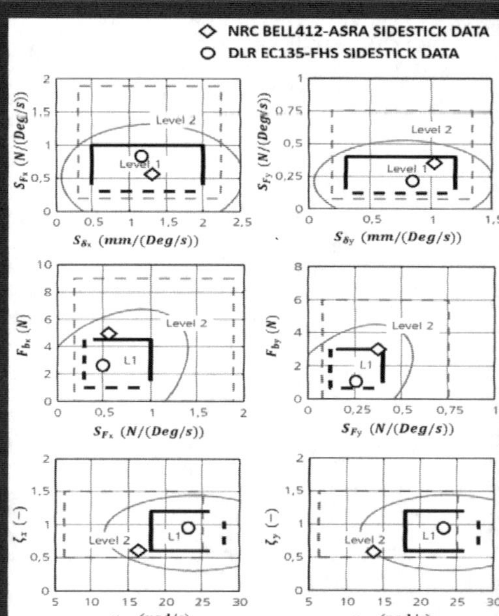

Closing Remarks

Summary and Conclusions

- Attitude Capture Study: Changes in workload (compensation, activity, strategy) with stick displacement-force limits.
- Hover Stabilization Results: Asymmetric human interface (anthropomorphic, ergonomic) requirements as well as the interrelationship between stick parameters and 'precision' low speed mission task performance.
- Mission Task Results: Influence of task cut-off frequency relative to task and control bandwidth as key stick configuration factors.
- Comparative Results: Implications of system architecture, control augmentation and aeromechanics characteristics on systems integration and tuning processes.

Acknowledgements

'Significant contributions made by *NRC Flight Research Laboratory* as well as 'l'Ecole du personnel navigant d'essais et de réception (EPNER; Istres, France)' are gratefully acknowledged.'

Questions

Contact: Marc Alexander, MarcDavid.Alexander@nrc-cnrc.gc.ca

NEAR EARTH AUTONOMY

Captain: An Architecture for Accreditable Autonomous Rotorcraft

VFS, 2025 Handling Qualities Meeting
Jan 22,23 2025 - Huntsville AL
Lyle Chamberlain, CTO Near Earth Autonomy
lyle.chamberlain@nearearth.aero

Selected Vehicle Integrations

Art of the Possible | **Pursuing Practical Applications with Accredited Systems**

Year	Description	Program
'06	*Flying fast & low near buildings, trees, and wires*	***DARPA OAV II***
'10	*Safe flight and safe landing on full-scale helicopter*	***ARMY COMBAT MEDIC***
'12 – '17	*Autonomous aggressive flight, landing at unprepared sites*	***ONR AACUS***
'17	*Urban air mobility (A3 Vahana)*	***COMMERCIAL***
'18	*Military cargo delivery at small and medium scales (TRV Series)*	***ARMY/USMC JCTD***
'19	*Long-range delivery of short-lived medical supplies (L3Harris FRV-90)*	***ARMY BLOOD DELIVERY***
'20	*Heavy cargo delivery (Kaman K-Max)*	***NAVAIR KMAX***
'21	*Commercial cargo delivery (VoloDrone)*	***COMMERCIAL***
'23	*Military cargo delivery (Kaman KARGO)*	***NAVAIR M-ULSA***
'24	*Military cargo delivery (H-60)*	***ARMY UNMANNED BLACKHAWK***

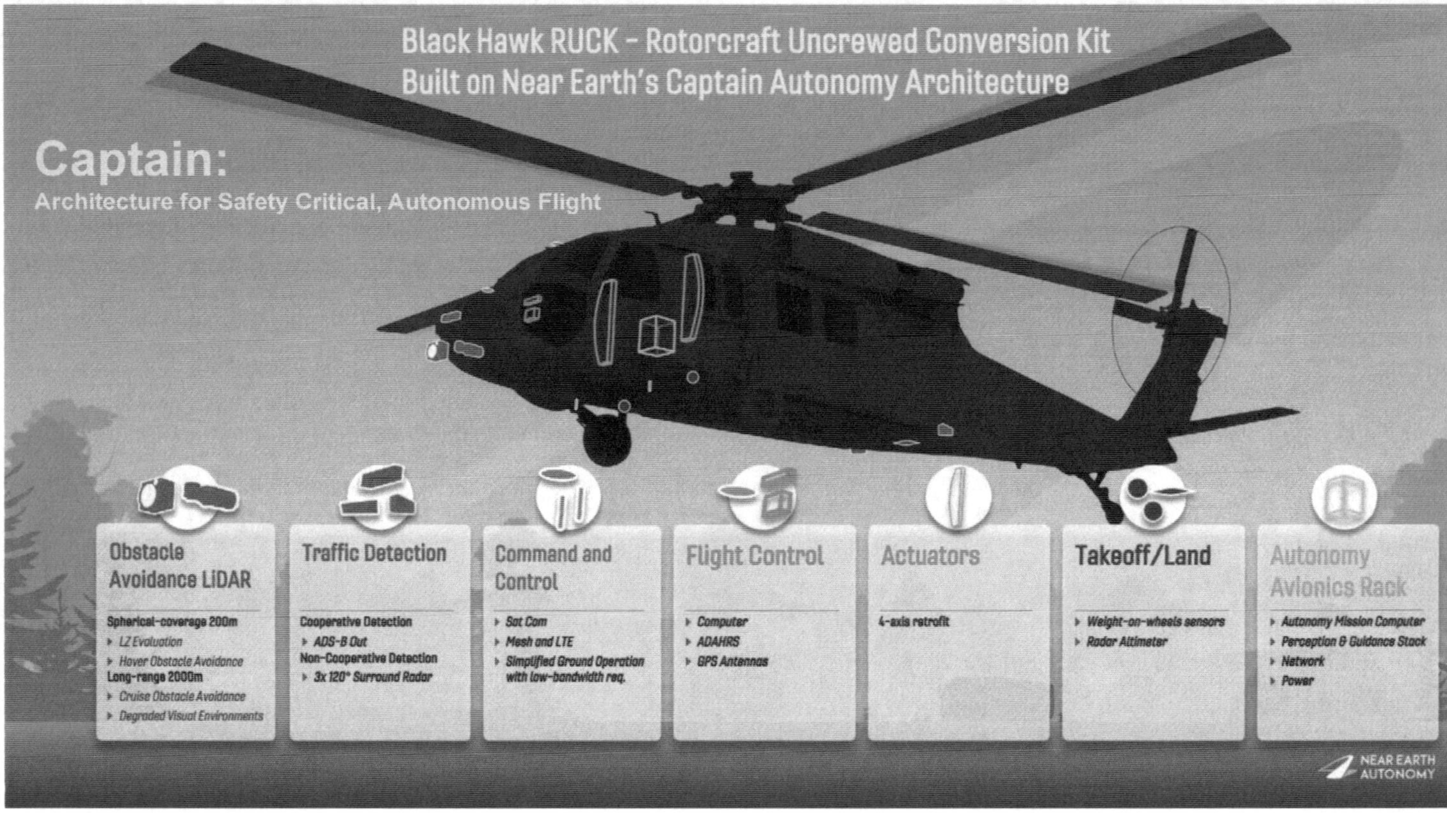

Black Hawk RUCK - Rotorcraft Uncrewed Conversion Kit
Built on Near Earth's Captain Autonomy Architecture
Captain:
Architecture for Safety Critical, Autonomous Flight
Obstacle Avoidance LiDAR
Spherical-coverage 200m
LZ Evaluation
Hover Obstacle Avoidance
Long-range 2000m
Cruise Obstacle Avoidance
Degraded Visual Environments
Traffic Detection
Cooperative Detection
ADS-B Out
Non-Cooperative Detection
3x 120° Surround Radar
Command and Control
Sat Com
Mesh and LTE
Simplified Ground Operation with low-bandwidth req.
Flight Control
Computer
ADAHRS
GPS Antennas
Actuators
4-axis retrofit
Takeoff/Land
Weight-on-wheels sensors
Radar Altimeter
Autonomy Avionics Rack
Autonomy Mission Computer
Perception & Guidance Stack
Network
Power
NEAR EARTH AUTONOMY

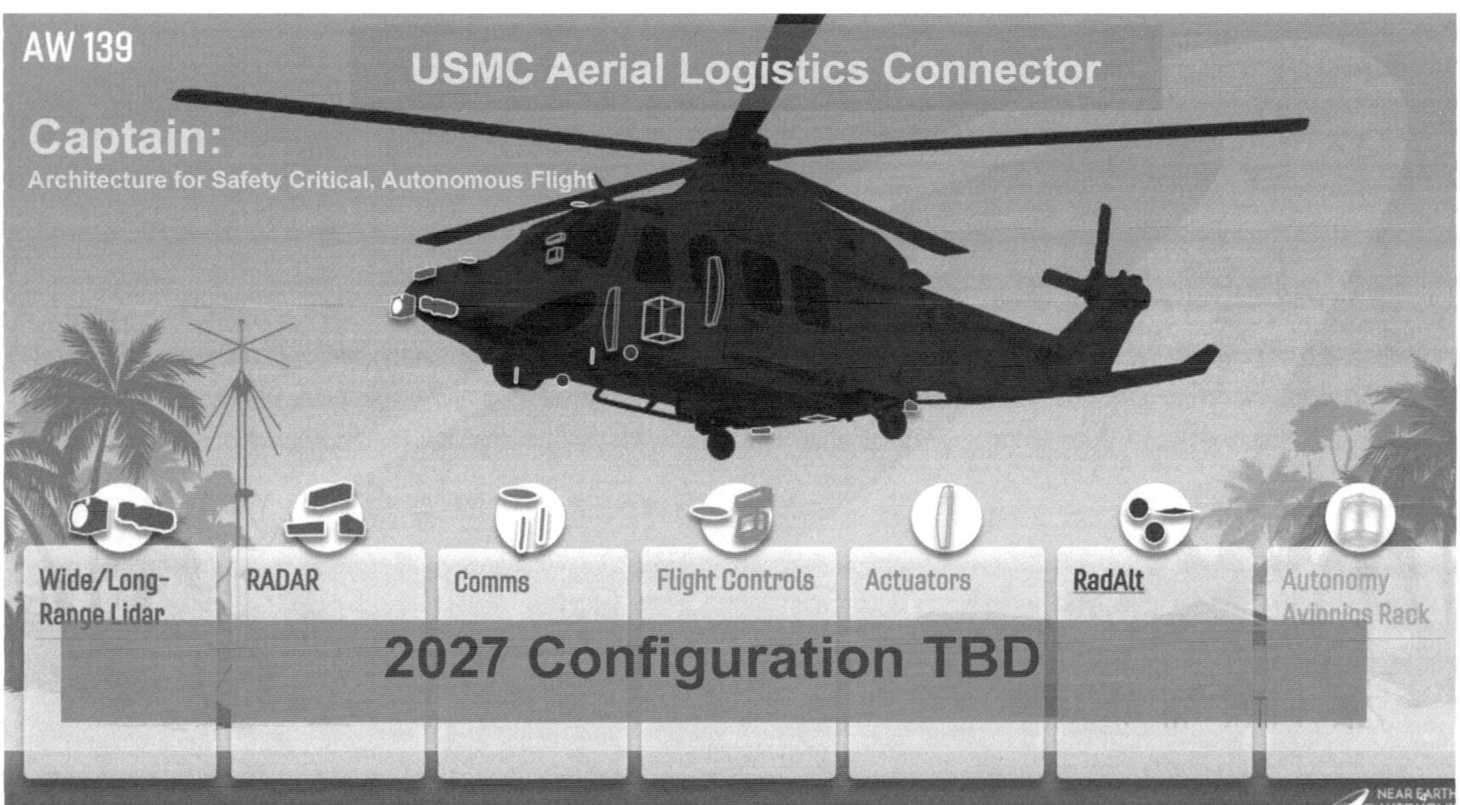

AW 139
USMC Aerial Logistics Connector
Captain:
Architecture for Safety Critical, Autonomous Flight
Wide/Long-Range Lidar
RADAR
Comms
Flight Controls
Actuators
RadAlt
Autonomy Avionics Rack
2027 Configuration TBD
NEAR EARTH AUTONOMY

Levels of Autonomy - Targets for Delivery/Logistics Conops

Autonomy Levels For Unmanned Systems (ALFUS)

REFERENCE LEVELS	MC Mission Complexity	EC Environmental Complexity	HI Human Independence
10	highest adaptation, decision space, team-of-teams collaborative missions; fully real-time planning; omniscient, highest level fidelity SA; human level performance lowest solution/possibility ratio	lowest margin for error, understandability; highest level of dynamics, variation, risks, uncertainty, mechanical constraints	performing on its own and approaching zero human interaction, negotiating with appropriate individuals
9 8 7	high adaptation, decision space, team collaborative missions/tasks; high real-time planning; strategic level, high fidelity SA	low solution/possibility ratio, understandability highly dynamic, complex, adversarial high risks, uncertainty, constraints	UMS informs humans; human provides strategic goals, interacting time between 6% and 35%
6 5 4	limited adaptation, decision space, vehicle tasking; limited real-time planning; tactical level, mid fidelity SA	mid solution/possibility ratio, understandability dynamic, simple mid risks, uncertainty, constraints	human approves decisions, provides tactical goals, interacting time between 36% and 65%
3 2 1	subsystem tasks/skills; internal, low fidelity SA	high solution/possibility ratio, understandability static, simple low risks, uncertainty, constraints	human decides, provides waypoints, interacting time between 66% & 95%
0	simplest, binary tasks	static, simple	remote control

DOD UAV Roadmap levels

#	Level
10	Fully Autonomous Swarms
9	Group Strategic Goals
8	Distributed Control
7	Group Tactical Goals
6	Group Tactical Replan
5	Group Coordination
4	Onboard Route Replan
3	Adapt to Failures & Flight Conditions
2	Real Time Health/Diagnosis
1	Remotely Guided Remotely Guide

Captain - Certifiable Autonomy Architecture

Autonomy Core:
Certified HW+SW
DAL A or Equivalent

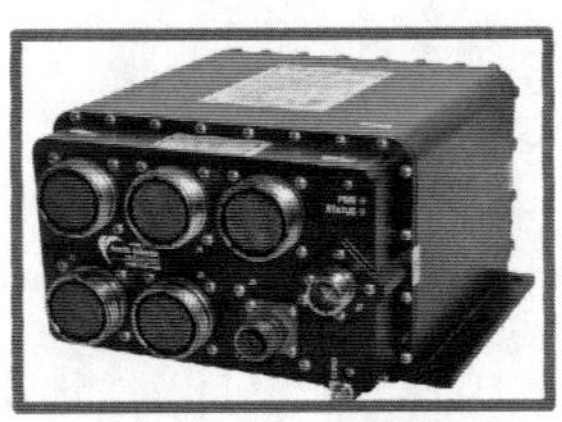

Autonomy Certification Target for Near Earth

Certified Core Functions:
- Mission Autonomy
- Autonomy Executive
- Runtime Assurance

Autonomy functions of varying assurance levels are tied together by a certifiable core system.

Aircraft Safety Case built around certifiable core system.

	Functions (other DAL)
	Perception Planning
	AVO Station
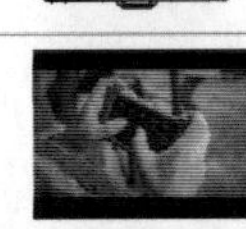	FO Tablet
	Comms
	Flight Control

Captain: Autonomy Architecture

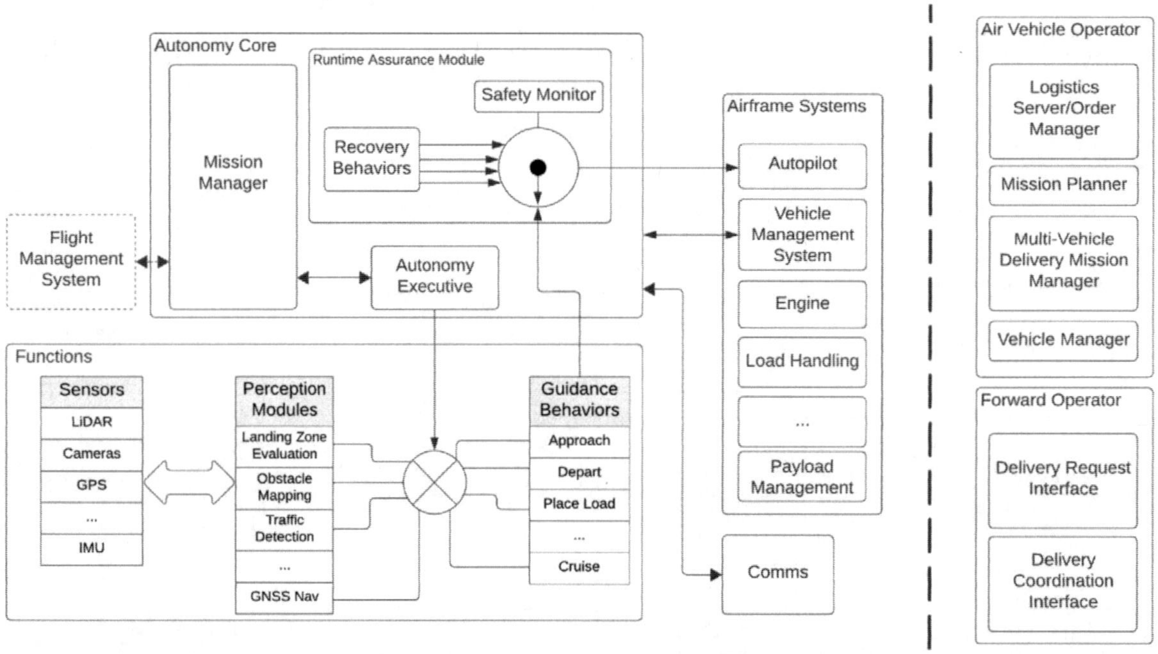

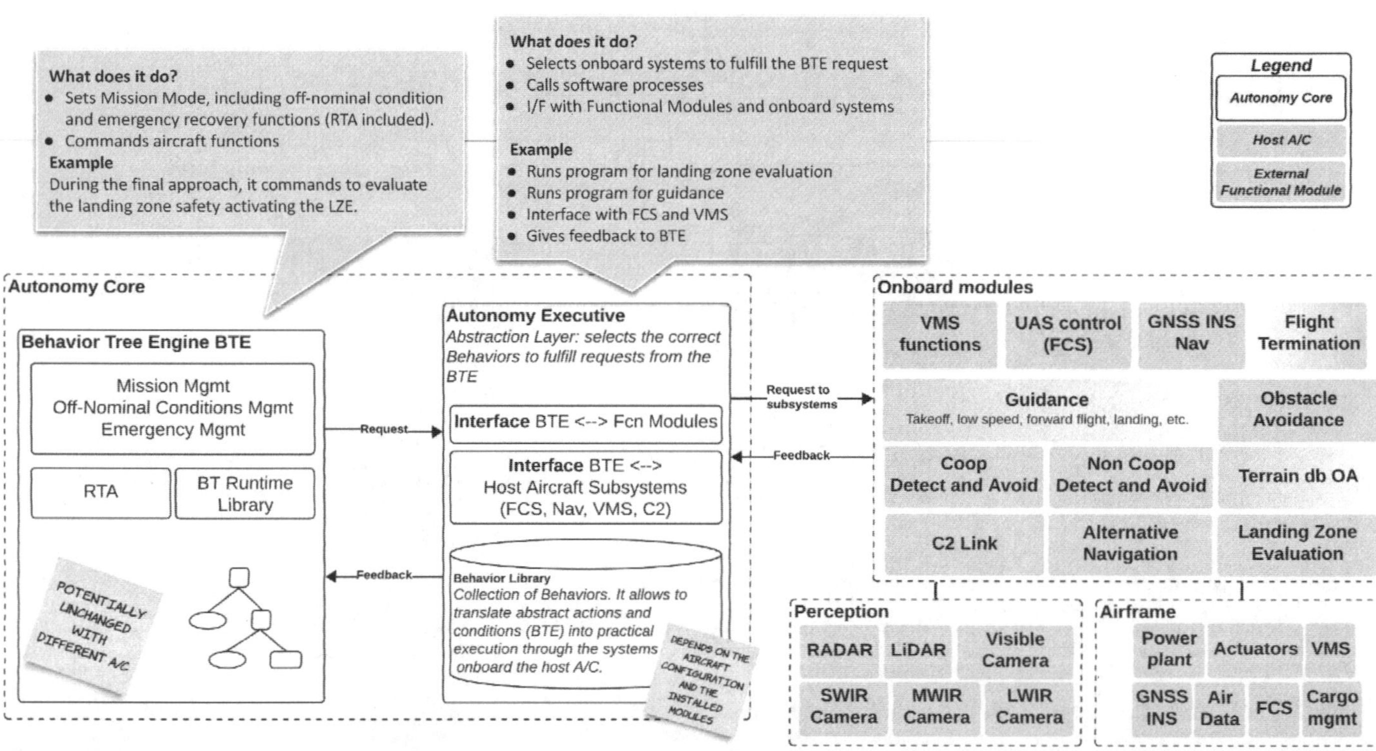

Migration Roadmap from Federated to Integrated Modular Avionics

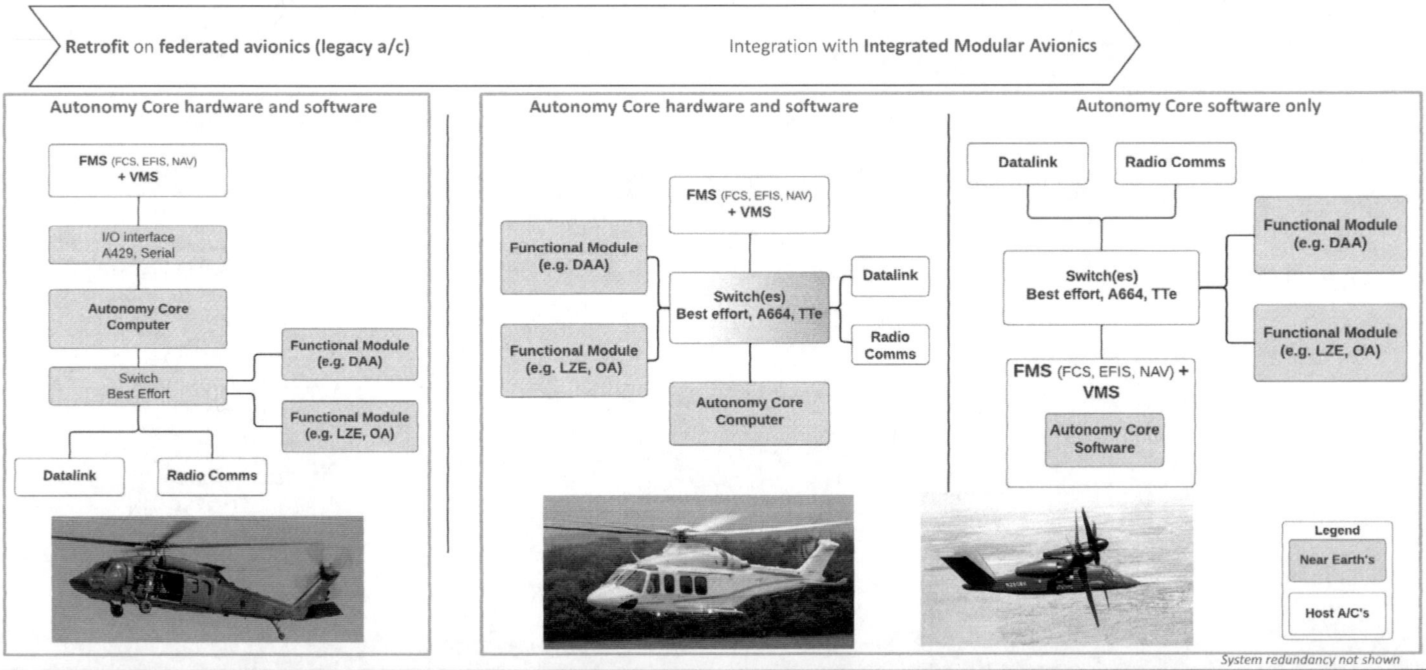

System redundancy not shown

NEAR EARTH AUTONOMY

For General Release | 1/22/2025 | 9

Hybrid Control and Guidance for Compatibility with Autonomous Decision Making

NEAR EARTH AUTONOMY

Recap: *Captain: Autonomy Architecture*

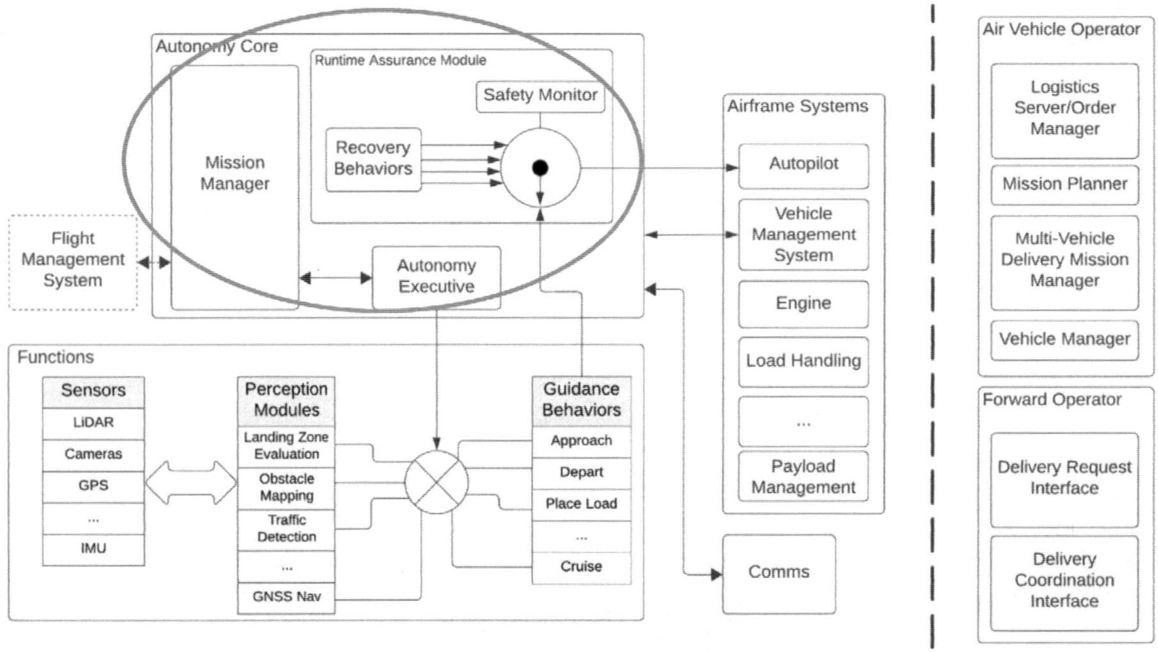

Model Checking of Mission Autonomy Logic

Near Earth Automated Model Checking Tools:
Automatically evaluate Mission Autonomy models for correctness using temporal logic:
- e.g. Safety, Liveness, Reachability, Invariance

Pipeline for specification & verification of Mission Autonomy logic

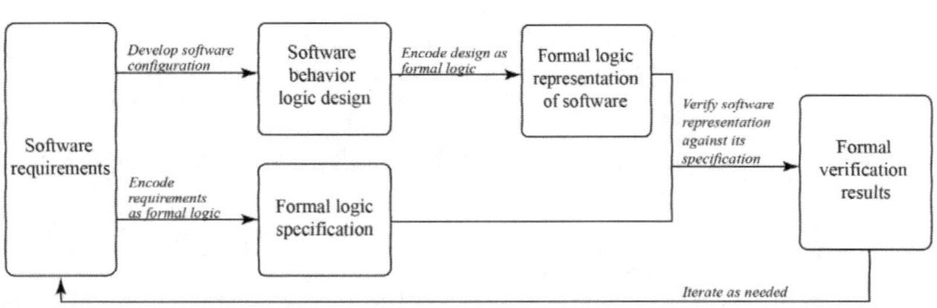

Resulting, strongly verified Mission Autonomy logic

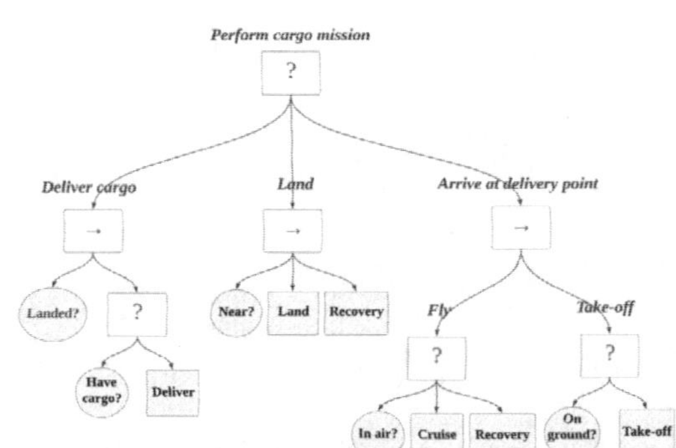

Recap: *Captain: Autonomy Architecture*

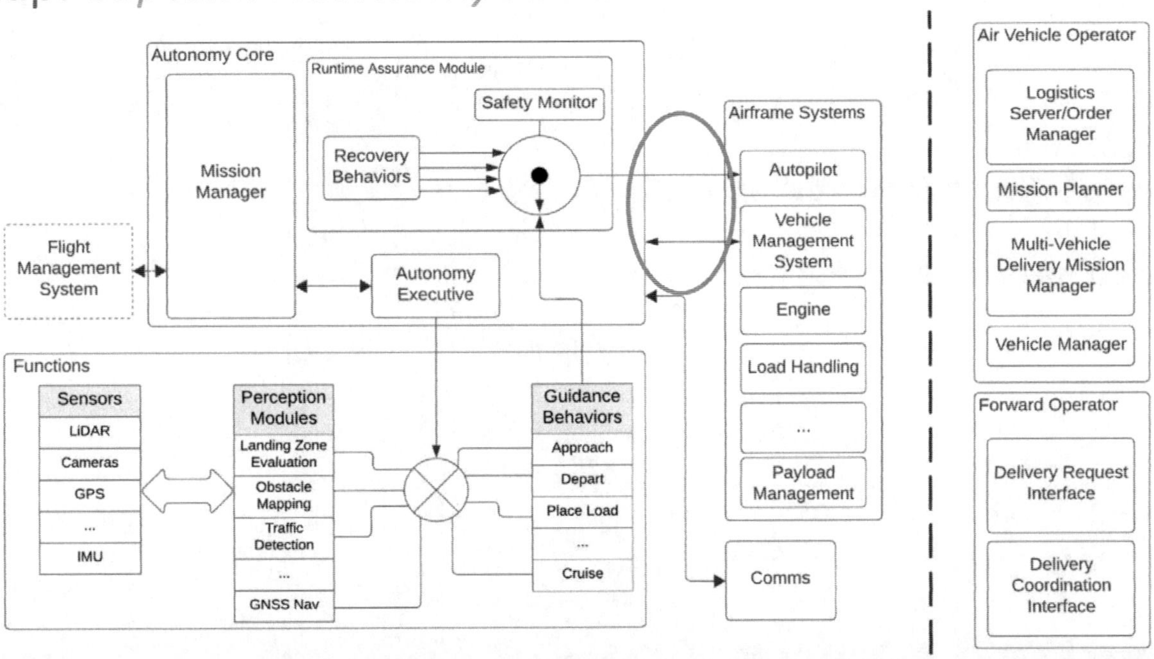

Example Captain Interfaces for Vehicle Guidance

FCS Guidance Commands

Hover / Low airspeed
- Ground-referenced velocity
- Yaw rate

Hover Hold
- Position Adjustment
- Heading Adjustment

Forward Flight
- Airspeed
- Bank angle
- Climb/descent rate

Stereotyped Primitives
- Touchdown from hover hold
- Liftoff to hover hold
- Autorotation entry
- ...

VMS Interfaces

Modes
- Engine Start
- Safety Aircraft Request
- Autopilot Enable Request
- ...

Aircraft State and Status
- INS state / status
- Engine Torque
- Airspeed

- Control mode
- ...

Safety and Robustness of Hybrid Control Switching

Safe Departure Sequence:
Multiple guidance policies with overlapping "regions of attraction" ensure safe envelope and robust control profile.

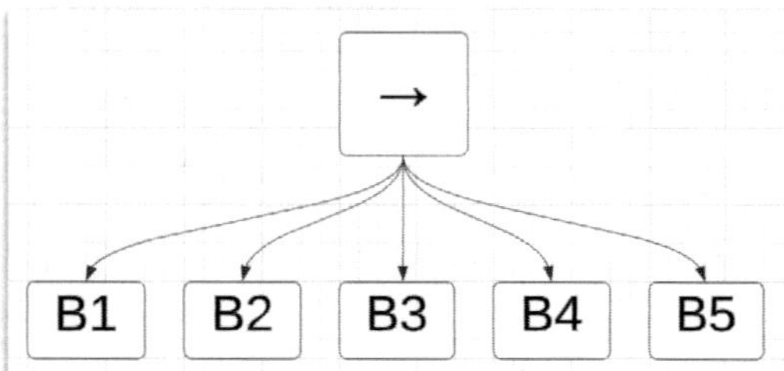

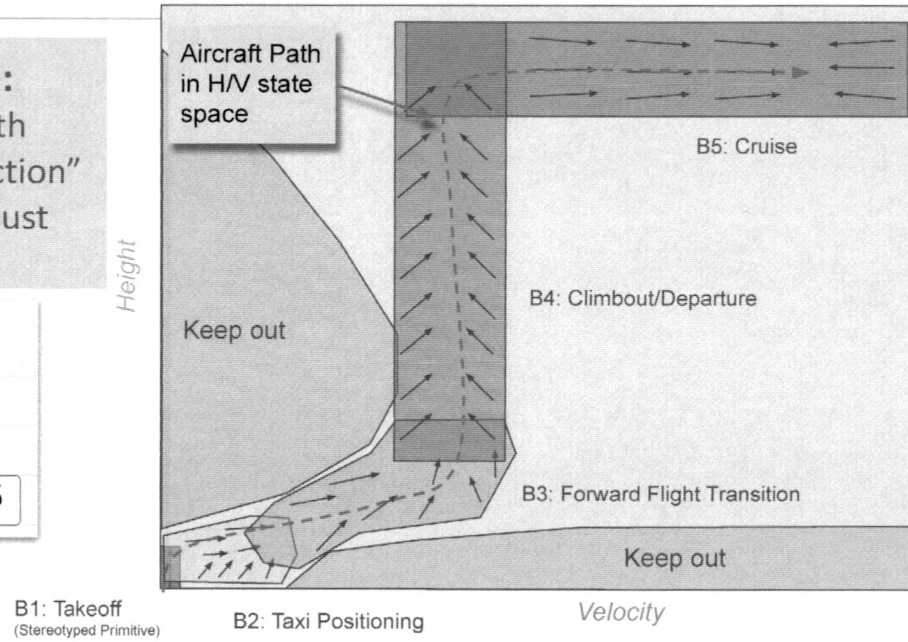

NEAR EARTH AUTONOMY

U.S. ARMY

Agenda

- About me.
- Human Factors
- Different Types of "Safety"
- Non-Quantified System Safety: Software System Safety Practices
- An Objective-Based Human Factors System Safety Approach: Strategy
- HFE scope derived from system functions
- Example

UNCLASSIFIED

Human Factors

- Human Factors in the field of safety is interested in behavior (errors, decision making, timely execution)
- This differs from other HF disciplines that focus on different objectives
  - User Acceptance
  - Usability and Safety/airworthiness

Different views on the relationship between usability and user experience: (a) broad-scope definition of user experience, (b) medium-scope definition of user experience, (c) narrow-cope definition of user experience
From: Sauer, J., Sonderegger, A., & Schmutz, S. (2020). Usability, user experience and accessibility: towards an integrative model. *Ergonomics, 63*(10), 1207-1220.

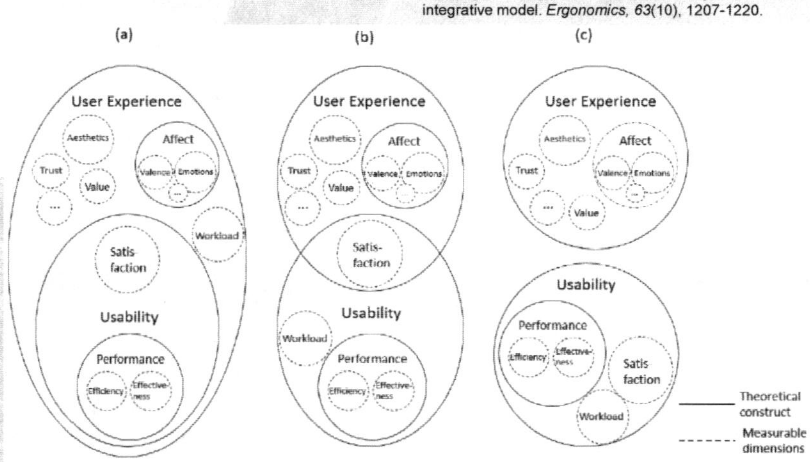

UNCLASSIFIED

UNCLASSIFIED

Types of Requirements

Design

- Specify a design attribute, function, or characteristic of the system that must be present.
- Example: "Each page of options (menu) shall have a title that clearly identifies the purpose of that menu" – 5.1.3.12.2.2.4 in MIL-STD-1472H

Process

- A process requirement is…a defined methodology with established techniques to fulfill the requirement.
- A process requirement is not…a recommendation to "collaborate" or "work with" other personnel.
- Example: SAE 4761 (the safety process as owned by the S-18 working group)

UNCLASSIFIED

U.S. ARMY

Different Types of "Safety"

System Safety

1. Identify hazards through the life cycle
2. Applicable to materiel solutions
3. Responsibility of the manufacturer
4. Based on the designed intended use

Operational Safety

1. Prevent accidents when using the system
2. Analysis specific to the environment and users
3. Responsibility of the user (organization or individual)
4. Based on the current use

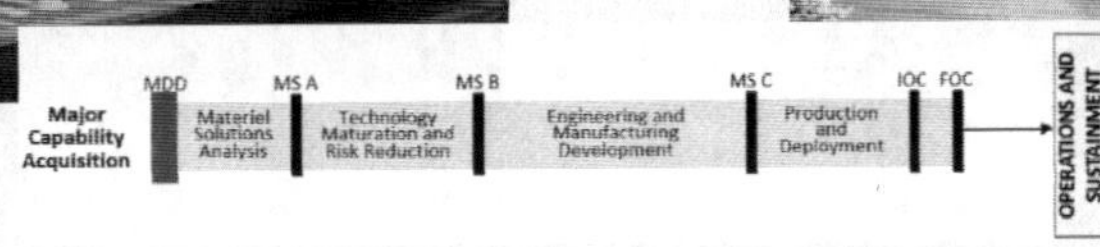

UNCLASSIFIED

U.S. ARMY

HFE scope derived from system functions

ARP 4761 Process

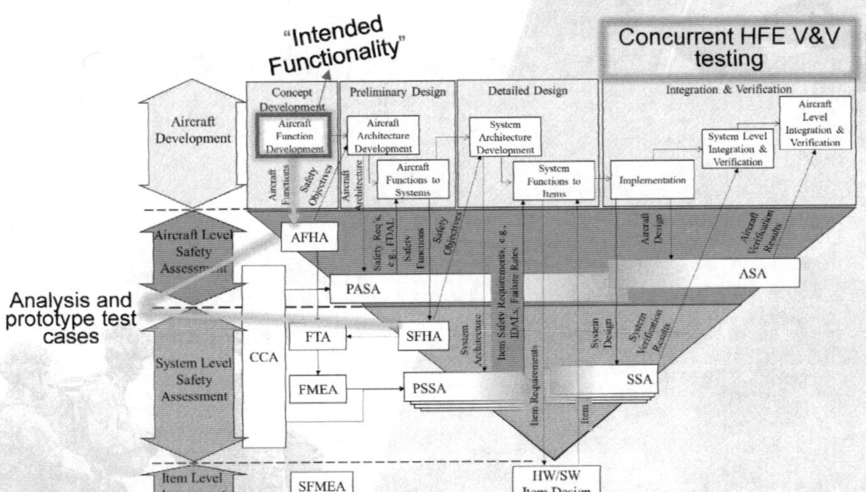

Carter, Glenn, Victor Terres, Chris Vinegar, and Jason Rupert. "Evaluating Army Airworthiness Safety Assessment Techniques for Autonomy, Artificial Intelligence, and Machine Learning." Paper presented at the International System Safety Society Conference, Redstone Arsenal, AL, 2024.

UNCLASSIFIED

UNCLASSIFIED

U.S. ARMY

Example Flight Controls FHA

Name	Documentation	Severity	Aircraft Effect	Crew Effect	Remarks
Malfunction of Flight Director Position Hold	"The aircraft is in position hold mode, but the system is driving the aircraft to an erroneous position. Flight crew will notice significant deviation, especially any flight path that trends towards another aircraft or terrain. After recognition, the flight crew will be able to disengage the mode to mitigate the effects of the erroneous commands and maintain continued safe flight and landing."	3 - Major	Significant Reduction in Aircraft Functional Capabilitie s or Safety Margins	Significant Increase in Crew Workload	"It is assumed that crew will have ability to override erroneous position commands and/or regain manual control of aircraft. It is assumed that the crew will recognize deviation in aircraft position."

UNCLASSIFIED

UNCLASSIFIED

U.S. ARMY

Example Flight Controls FHA

Name	Documentation	Severity	Aircraft Effect	Crew Effect	Remarks
Malfunction of Flight Director Position Hold	"The aircraft is in position hold mode, but the system is driving the aircraft to an erroneous position. Flight crew will notice significant deviation, especially any flight path that trends towards another aircraft or terrain. After recognition, the flight crew will be able to disengage the mode to mitigate the effects of the erroneous commands and maintain continued safe flight and landing."	3 - Majo	Capabilitie s or Safety Margins		"It is assumed that crew will have ability to override erroneous position commands and/or regain manual control of aircraft. It is assumed that the crew will recognize deviation in aircraft position."

Assumptions treated as requirements for Validation/Verification activities

Condition treated as the test scenario

UNCLASSIFIED

U.S. ARMY

Non-Quantified System Safety: Software System Safety Practices

- The goal is to identify risk, as defined by a quantified then categorized probability and categorized severity
- Hardware: Probability defined in terms of predicted time to failure
- Software: Probability assessed in terms of "software contribution to system risk…that considers the potential risk severity and the degree of control that software exercises over the hardware."
  - Probability is assessed using level of rigor tasks, or "analysis and verification activities necessary to provide a sufficient level of confident that a safety-critical or safety-related software function will perform as required."
  - Level of rigor tasks are defined based on the combination of level of control the software has over the function and the severity of the function's failure

RISK ASSESSMENT MATRIX

PROBABILITY \ SEVERITY	Catastrophic (1)	Critical (2)	Marginal (3)	Negligible (4)
Frequent (A)	High	High	Serious	Medium
Probable (B)	High	High	Serious	Medium
Occasional (C)	High	Serious	Medium	Low
Remote (D)	Serious	Medium	Medium	Low
Improbable (E)	Medium	Medium	Medium	Low
Eliminated (F)	Eliminated			

SwCI	Level of Rigor Tasks
SwCI 1	Program shall perform analysis of requirements, architecture, design, and code; and conduct in-depth safety-specific testing.
SwCI 2	Program shall perform analysis of requirements, architecture, and design; and conduct in-depth safety-specific testing.
SwCI 3	Program shall perform analysis of requirements and architecture, and conduct in-depth safety-specific testing.
SwCI 4	Program shall conduct safety-specific testing.
SwCI 5	Once assessed by safety engineering as Not Safety, then no safety specific analysis or verification is required

U.S. ARMY

Non-Quantified System Safety: Software System Safety Practices

- The goal is to identify risk, as defined by a quantified then categorized probability and categorized severity
- Hardware: Probability defined in terms of predicted time to failure
- Software: Probability assessed using level of rigor tasks.
- Human Factors hazards need to be assessed under this framework of severity and probability.
  - We can define severity of an action, the same as other functional failure.
  - Human error attributed to design is difficult to quantify, though.

RISK ASSESSMENT MATRIX

PROBABILITY \ SEVERITY	Catastrophic (1)	Critical (2)	Marginal (3)	Negligible (4)
Frequent (A)	High	High	Serious	Medium
Probable (B)	High	High	Serious	Medium
Occasional (C)	High	Serious	Medium	Low
Remote (D)	Serious	Medium	Medium	Low
Improbable (E)	Medium	Medium	Medium	Low
Eliminated (F)	Eliminated			

SwCI	Level of Rigor Tasks
SwCI 1	Program shall perform analysis of requirements, architecture, design, and code; and conduct in-depth safety-specific testing.
SwCI 2	Program shall perform analysis of requirements, architecture, and design; and conduct in-depth safety-specific testing.
SwCI 3	Program shall perform analysis of requirements and architecture, and conduct in-depth safety-specific testing.
SwCI 4	Program shall conduct safety-specific testing.
SwCI 5	Once assessed by safety engineering as Not Safety, then no safety specific analysis or verification is required

U.S. ARMY

An Objective-Based Human Factors System Safety Approach: Strategy

Engineering standard consisting of 3 elements:

1. Level of rigor classifications based on severity of failure and level of user control
   - Classifications derived from system functional hazard analysis (FHA) (the same as software)
   - Define the scope of work necessary for confidence of optimal human performance (reduce requirements creep)
   - Based on levels of automation and severity of failure
2. Audit checkpoints and defined CDRL interfaces with other processes which can be traced to the phases of the acquisition life cycle
   - Fill the concern of prototype user testing and potential pathway to box-level Human Factors qualification
   - Information interfaces are defined
3. Guidance for assessing human error contribution to system safety
   - Human Factors hazards are historically difficult to characterize (i.e., characterizing the hazard of activating the incorrect switch in a sequence)
   - Pre-established guidance increases standardization and acknowledges that Human Factors hazard contribute to overall system risk

* Proposed new HFE standard of practice

U.S. ARMY

An Objective-Based Human Factors System Safety Approach: Strategy

1. Level of rigor classifications based on severity of failure and level of user control
   - Classifications derived from system functional hazard analysis (FHA) (the same as software) (Already required and done for all systems – new or modifications)
   - Based on levels of automation and severity of failure
   - Define the scope of work necessary for confidence of optimal human performance (reduce requirements creep) (Utilize existing techniques, but the library of tools and categorization of work to required confidence needs to be defined)
     - Example: Define lower-level design requirement (i.e., MIL-STD-1472).
   - Program then completes the predefined list of analyses and testing for delivery of certification

Function	Failure ID	Phases	Failure Condition	Failure Effect	Failure Classification	Justification	Qualitative Probability	Remarks
Provide WiFi on board	46_40_01_01	All	Provide WiFi when not required	WiFi signal will be active without any action from the flight crew	No safety affect	This FC has neither an effect on the operation capability nor on the crew workload	Remote	Though not having any effect on safety, this FC could lead to a non-controlled security leak
	46_40_01_02	All	Loss of WiFi	WiFi signal will not be available to any PED	No safety affect	This FC has neither an effect on the operation capability nor on the crew workload. It could cause some discomfort on the passengers	Probable	

Example of a simple, industry FHA from: https://dmd.solutions/blog/2023/12/19/what-is-functional-hazard-assessment/

U.S. ARMY

An Objective-Based Human Factors System Safety Approach: Strategy

2. Audit checkpoints and defined CDRL interfaces with other processes which can be traced to the phases of the acquisition life cycle
   - Fill the concern of prototype user testing and potential pathway to box-level Human Factors qualification
   - Define what information is needed from Human Factors and what information is passed to Human Factors (i.e., S-18H document recently reviewed)
     - Working Groups and collaboration is great, but ensuring safety of our operators should come from a repeatable process that can be easily communicated, outlined, and tracked. *Good people is not a resilient process.*

U.S. ARMY

An Objective-Based Human Factors System Safety Approach: Strategy

3. Guidance for assessing human error contribution to system safety
   - Human Factors hazards are historically difficult to characterize (i.e., characterizing the probability of activating the incorrect switch in a sequence)
   - Pre-established guidance increases standardization and acknowledges that Human Factors hazard contribute to overall system risk

RELATIONSHIP BETWEEN SwCI, RISK LEVEL, LOR Tasks, AND RISK

Software Criticality Index (SwCI)	Risk Level	Software LOR Tasks and Risk Assessment/Acceptance
SwCI 1	High	• If SwCI 1 LOR tasks are unspecified or incomplete, the contributions to system risk will be documented as HIGH and provided to the PM for decision. The PM shall document the decision of whether to expend the resources required to implement SwCI 1 LOR tasks or prepare a formal risk assessment for acceptance of a HIGH risk.
SwCI 2	Serious	• If SwCI 2 LOR tasks are unspecified or incomplete, the contributions to system risk will be documented as SERIOUS and provided to the PM for decision. The PM shall document the decision of whether to expend the resources required to implement SwCI 2 LOR tasks or prepare a formal risk assessment for acceptance of a SERIOUS risk.
SwCI 3	Medium	• If SwCI 3 LOR tasks are unspecified or incomplete, the contributions to system risk will be documented as MEDIUM and provided to the PM for decision. The PM shall document the decision of whether to expend the resources required to implement SwCI 3 LOR tasks or prepare a formal risk assessment for acceptance of a MEDIUM risk.
SwCI 4	Low	• If SwCI 4 LOR tasks are unspecified or incomplete, the contributions to system risk will be documented as LOW and provided to the PM for decision. The PM shall document the decision of whether to expend the resources required to implement SwCI 4 LOR tasks or prepare a formal risk assessment for acceptance of a LOW risk.
SwCI 5	Not Safety	• No safety-specific analyses or testing is required.

RISK ASSESSMENT MATRIX

SEVERITY / PROBABILITY	Catastrophic (1)	Critical (2)	Marginal (3)	Negligible (4)
Frequent (A)	High	High	Serious	Medium
Probable (B)	High	High	Serious	Medium
Occasional (C)	High	Serious	Medium	Low
Remote (D)	Serious	Medium	Medium	Low
Improbable (E)	Medium	Medium	Medium	Low
Eliminated (F)	Eliminated			

U.S. ARMY

Questions / Guidance

U.S. ARMY

Information flow

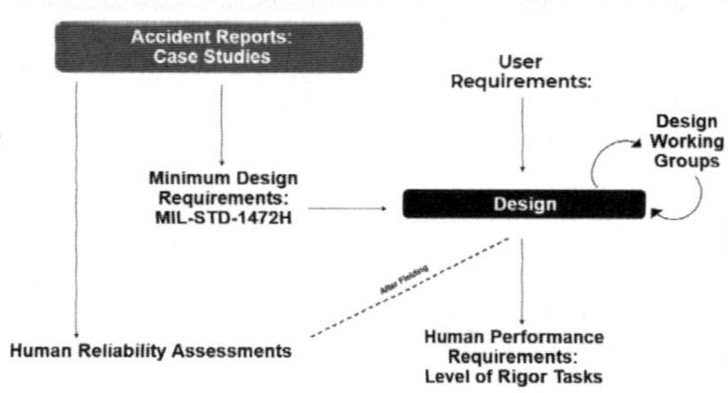

Concerns with utilizing only current design requirements

- Most of the current approaches assume information availability that is not realized until later in the program.
  - How can you specify design requirements if you don't know the technology to be implemented (i.e., touchscreen v. traditional screens)
- The analysis/testing requirements should be sourced from the design, not a task decomposition of legacy systems/task flow.
  - Assess ALL functionality, not just the functions users are aware of…
- Current HFE requirements rely on final testing to meet the airworthiness and program requirements. Without design reviews, programs have no method to "burn down" programmatic risk through the development process.

U.S. ARMY

Human Factors as a Practice

- Human Factors is the incorporation of the human in the "system" from cognition to physiological action to system performance
- What is the goal?
  - User Acceptance
  - Safety/airworthiness
- Currently, how do we assess?
  - Cognitive level – self-report by user *Easier and correlates with system performance on traditional systems
  - System level – performance of task
  - All targeted at "critical tasks" as defined by legacy systems and systems experts
- But the human's role is changing, now what?
  - The correlation work on self-report metrics to system performance is based on traditional systems, so we'll likely see more performance-based testing to accommodate surveys (e.g., workload)
  - Critical tasks are difficult to identify for new systems

Leonardo Helicopters

Handling Qualities MTEs Evaluation Methodologies in Preliminary Design of High-Speed Rotorcraft Configurations

F. Riccardi, M. Mamino, M. Pecoraro*, A. Ragazzi

VFS Handling Qualities Technical Meeting
Jan. 22-23, 2025 – Huntsville, AL, USA

SUMMARY

- Introduction & Motivations
- Developed MTEs Preliminary Evaluation Methodology
- Fast Rotorcraft Comparison executing MTEs
- MTEs performance sensitivity analysis to design parameters (focus on tiltrotor)
- Concluding Remarks & Way Forward

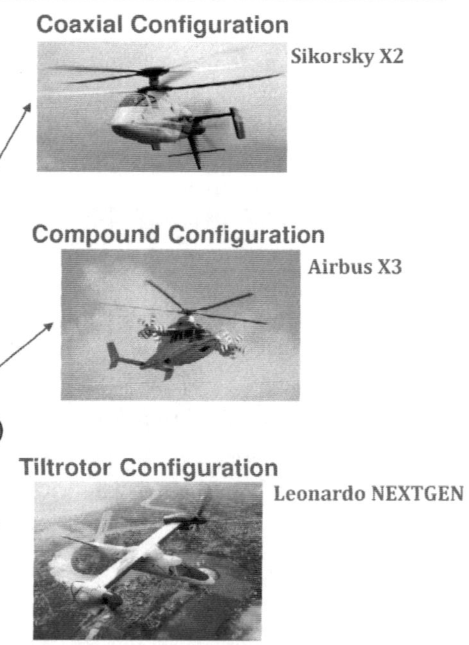

Introduction & Motivations

Rotorcraft HQs evaluation have been usually neglected in conceptual and early preliminary design processes because at these early stages not enough details are available to perform an assessment.

- **State of the art:** Lawrence et al. [1] and Zanoni et al. [2] both incorporated HQs into a NDARC-based conceptual design process, adopting different approach to SCAS preliminary synthesis.
  ADS-33 HQs metrics assessed and enforced in NDARC loop are: short term (bandwidth vs. time delay), moderate amplitude (quickness criteria) and large amplitude response (control power).
  Once design iterations are concluded, MTEs execution can be done by means of piloted simulations.

- **Need:** Enforce at early design stage a preliminary HQs evaluation directly in terms of key MTEs performance via fast desktop simulations (without real-time piloted sessions), to feedback design and highlight peculiarities on fast rotorcraft architectures: coaxial lift offset compound helicopter with pusher propeller in tail (COAX), single main rotor compound with two lateral propellers tip mounted on box wing (COWP) and tiltrotor (TILT).

- **Approach:** A simplified approach is developed, based on reduced number of bare A/C stability and control derivatives from LTI models (fidelity compatible with design phase), including a preliminary control loop closure and proper piloting logic to perform MTEs of interest (in low and high-speed regime).

[1] Lawrence, B., Theodore, C.R., Johnson, W., Berger, T., "A handling qualities analysis tool for rotorcraft conceptual designs", The Aeronautical Journal, 122(1252), 960–987 (2018)
[2] Zanoni, A., Gerosa, G., Di Lallo, L., Masarati, P., "Handling Qualities in Rotorcraft Conceptual Design", Aerotecnica Missili & Spazio, 101, 95–108 (2022)

Considered Fast Rotorcraft Architectures

Gross Weight Medium-Heavy Class Rotorcraft

Generic Coaxial Configuration (COAX)

Generic Winged Single Main Rotor Propellers Thrust Compound (COWP)

Generic Tiltrotor Configuration (TILT)

Axis	Low Speed	High Speed	Low Speed	High Speed	Low Speed	High Speed
Pitch	Rotors Long. Cyclic + Propeller Col.	Elevator + Rotors Long. Cyclic + Propeller Col.	MR Long. Cyclic	Elevator + MR Long. Cyclic	Prop-Rotors Sym. Long. Cyclic + Sym. Nacelle Tilt	Ruddervator Sym. + Prop-Rotors Sym. Long. Cyclic
Roll	Rotors Lat. Cyclic	Rotors Lat. Cyclic	MR Lat. Cyclic + Flaperon Diff.	Flaperon Diff. + MR Lat. Cyclic	Prop-Rotors Diff. Col. + Lat. Cyclic	Flaperon Diff. + Prop-Rotors Diff. Long. Cyclic
Yaw	Diff. Col. Rotors	Rudder	Diff. Col. Propellers	Rudder	Prop-Rotors Diff. Long. Cyclic + Diff. Nacelle Tilt	Ruddervator Diff. + Prop-Rotors Diff. Col.

Preliminary MTEs Evaluation Methodology – General overview

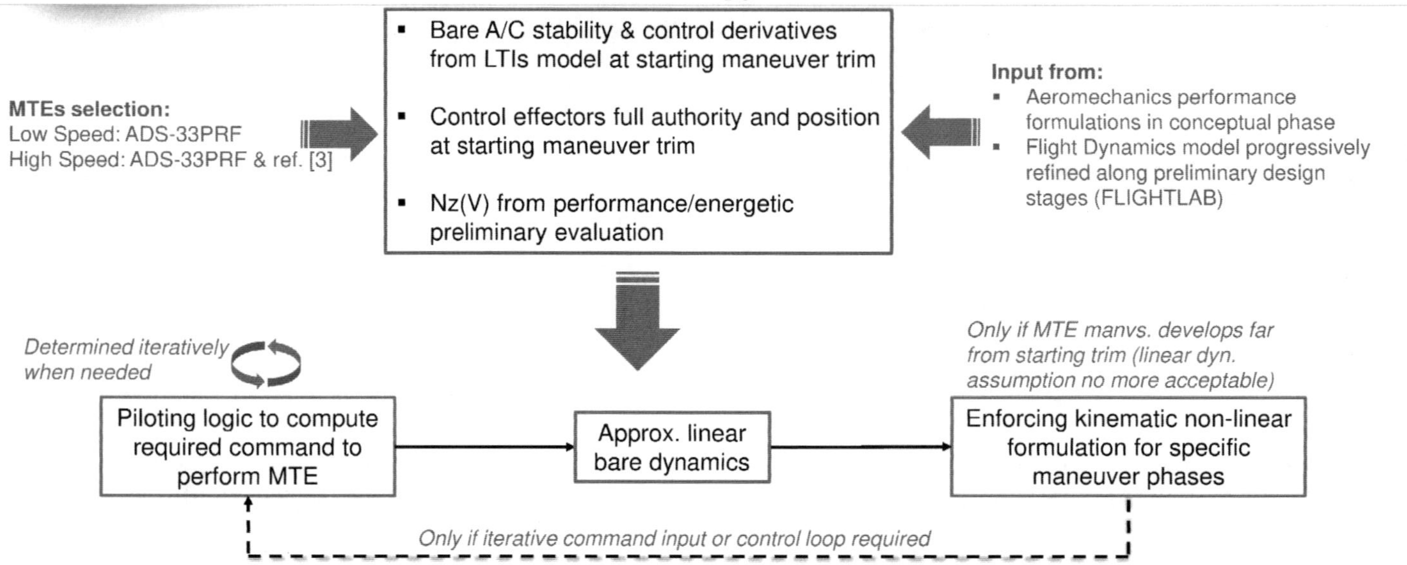

[3] Berger, T., "Handling Qualities Requirements and Control Design for High-Speed Rotorcraft" FCDD-AMV-20-01, 2020

Preliminary MTEs Evaluation Methodology – Low Speed cases overview

MTE	REF.	Initial Trim	Bare A/C approx. dynamics	Stability deriv./ RB modes data	Ctrl deriv.	Other input	Kinematic NL correction	Ctrl loop	Iterative command computing
Hovering Turn	ADS-33 3.11.4	Hover	1DoF 1st order from δped to r	$N_{/r}$	$N_{/\delta ped}$	-	NO	NO	YES
Lateral reposition	ADS-33 3.11.8	Hover	2DoF 3rd order from δlat to ϕ 1st order from ϕ to v	$Y_{/v}$ Roll subsidence: τ_r Dutch-roll: ω_{dr}, ζ_{dr}	$L_{/\delta lat}$	-	NO	YES	YES
Depart Abort	ADS-33 3.11.7	Hover	2DoF 4th order from δlon to θ 1st order from θ to u	$X_{/u}$,$Z_{/u}$,$Z_{/w}$ $M_{/w}$, $M_{/w}$, $M_{/q}$	$M_{/\delta lon}$ $X_{/\delta lon}$	-	NO	YES	YES
Vertical maneuver	ADS-33 3.11.6	Hover	1DoF 1st order from δcol to w	$Z_{/w}$	$Z_{/\delta col}$	Max/Min Nz	NO	NO	YES

Preliminary MTEs Evaluation Methodology – High Speed cases overview

MTE	REF.	Initial Trim	Bare A/C approx. dynamics	Stability deriv./ RB modes data	Ctrl deriv.	Other input	Kinematic NL correction	Ctrl loop	Iterative command computing
Acceleration /Deceleration	proposed in [3]	-	-	-	-	Max/Min Nx(V)	Given Nx(V) time integration to find Vx, X	NO	NO
Pull-Up /Push-Over (NOE)	ADS-33 3.11.15	Level flight V sweep	3DoF 2nd order from δlon to Nz 2nd order from δlon to q 2nd order from δlon to α	$Z_{/w}$, $Z_{/\alpha}$ Short period: ω_{sp}, ζ_{sp}	$M_{/\delta lon}$	Max/Min Nz(V)	NO	NO	NO
Break Turn	proposed in [3]	Level flight V sweep	1DoF 1st order from δlat to p [roll-in/out phases]	$L_{/p}$	$L_{/\delta lat}$	Max Nz(V)	YES [steady bank phase]	NO	YES

[3] Berger, T., "Handling Qualities Requirements and Control Design for High-Speed Rotorcraft" FCDD-AMV-20-01, 2020

General premises

- LHD has adopted an unbiased methodology to preliminarily estimate the MTEs performance of three different Fast Rotorcraft architectures under examination.
- This approach allowed to conduct an analysis of each of the three architectures with the same, comparable, analytical assumptions and level of detail, with the aim of ensuring a fair comparison.
- All relevant data in the study have been obtained from public available sources and/or LHD's Proprietary Data, ensuring therefore the best possible representativeness for each of the three architectures.
- The input used are valid to report a comparative methodology in early design phase. Results are depending on the aeromechanics models used, that can be further refined, to enhance the analysis precision and/or improve maneuvering capabilities.
- No considerations are enforced to consider possible rotors flapping limitation during maneuvering conditions.

Fast Rotorcraft Comparison executing MTEs

Included methodology implementation details for each MTE

Fast Rotorcraft Low Speed MTEs Preliminary Assessment Overview

		Criteria	COAX	COWP	TILT
Low Speed (ADS33)	Hovering Turn	Time to complete maneuver Desired < 10 s 10 s < Adequate < 15 s	**Desired** Coaxial MR high control derivative $N_{/\delta ped}$ combined with low I_{zz} inertia	**Desired** High damping derivative $N_{/r}$, asymmetric behavior due to propeller working cond.	**Desired** Adopted only diff. long. cyclic, performance improvement adding diff. nacelle tilt control
	Vertical Reposition	Time to complete maneuver Desired < 10 s 10 s < Adequate < 15 s	**Desired** High Nz max hovering thanks to high coaxial MR thrust margin	**Desired** Performance limited by max MR thrust available	**Desired** Performance limited by max MR thrust available
	Lateral Reposition Task Detailed	Time to complete maneuver Desired < 18 s 18 s < Adequate < 22 s	**Adequate** Performance limited for ensuring vertical equilibrium	**Adequate (L) Not Adequate (R)** Asymmetric behavior due to hover asymmetries	**Adequate** Adopted only diff. coll. pitch, performance improvement adding lat. cyclic control and increasing DCP authority
	Depart-Abort	Time to complete maneuver Desired < 25 s 25 s < Adequate < 30 s	**Desired** Adopted only long. cyclic, performance improvement adding propeller control	**Desired** Propeller near to saturation for anti-torque, not enough margin for longitudinal control	**Desired** Adopted only symm. long. cyclic, performance improvement adding symm. nacelle tilt control

Fast Rotorcraft Comparison – Low Speed: Lateral Reposition

ADS-33 3.11.8 Lateral Reposition

Start in a stabilized hover. Initiate a lateral acceleration to approximately 35 knots ground speed followed by a deceleration to laterally reposition the rotorcraft in a stabilized hover 400ft down the course within a specified time.

Time to complete

Cargo/Utility GVE = 18s (DESIRED) ; 22s (ADEQUATE)

Bare A/C approx. 2DoF dynamics

$$\frac{\phi}{\delta_{lat}}(s) = \frac{L_{/\delta lat}(s - Y_{/v})}{(s + {}^{1}/_{\tau_r})(s^2 + 2\zeta_{dr}\omega_{dr}s + \omega_{dr}{}^2)}$$

$$\frac{v}{\phi}(s) = \frac{g}{(s - Y_{/v})}$$

Unstable dutch-roll in hover, **control loop** needed

- PID on Φ loop (effector limits saturation)
- P on v loop
- P on y loop
- 2nd order shaping function on ϕ_{ref} from v loop

[applied default "pidtune" Matlab req.]

Piloting logic

Increase iteratively y_{ref} ramp slope until 400ft target is achieved with stable response.

Enforced $\phi_{max} = 30deg$ constraint in iterative method.

COWP asymmetric behavior: rolling moment to L generated by propeller asymmetric working regime

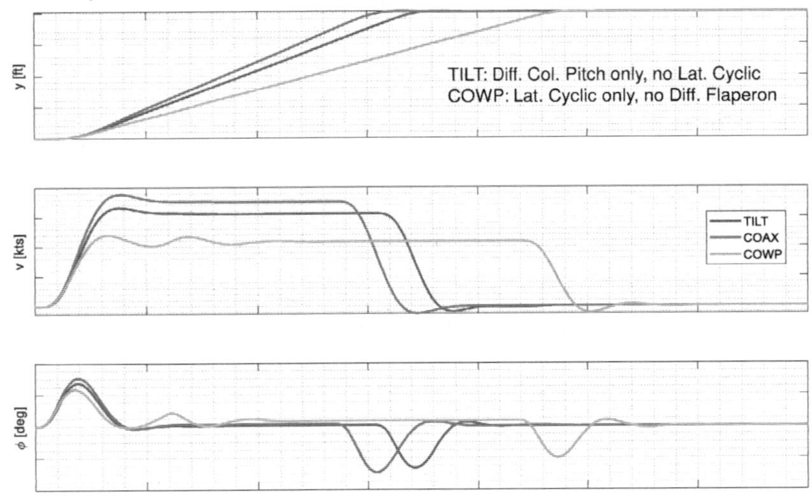

Fast Rotorcraft High Speed MTEs Preliminary Assessment Overview

		Criteria	COAX	COWP	TILT
Low to High Speed	**Acceleration / Deceleration 50-200-50kts**	Lowest Time to complete	**42s** Selected 200kts to have a fair comparison with TILT, above acceleration limited by power margin	**43s** Selected 200kts to have a fair comparison with TILT, above acceleration limited by power margin	**44s - 47s** Nacelle rate design increases overall performances, to be defined at early design phases to properly exploit power margin
High Speed	**Break Turn (180deg)**	Lowest Time to complete	**12s - 34s** Best in roll-in/out phases, coaxial MR high roll control power. Limited Nz (steady bank angle) at high speed by power margin	**14s - 55s** Worst in roll-in/out phases: asymmetric due to MR downwash on wing, high damping derivative $L_{/p}$. Limited Nz (steady bank angle) at high speed by power margin	**22s - 19s** Limited Nz (steady bank angle) at low speed wrt COAX-COWP, major benefit at high speed, due to wing lift
	Pull-Up/ Push-Over Task Detailed	Highest Average Ramp Angle	**9deg-1deg** Limited Nz at high speed by power margin	**7deg-1deg** Limited Nz at high speed by power margin	**5deg-6deg** Limited Nz at low speed by wing lift, design margin considering high lift devices

Results (and design space) does not target operative requirements and reflects generic fast rotorcraft architectures.

The methodology correctly represents the advantage for a winged architecture (e.g. the TILT) as the speed increases above typical conventional helicopter figures.

Fast Rotorcraft Comparison – High Speed: Pull-up/Push-Over NOE 1/2

ADS-33 3.11.15 Pull-up/Push-over

From level flight at the lesser of VH or 120 knots, attain a sustained positive load factor in a symmetrical pull-up. Transition, via a symmetrical push-over, to a sustained negative load factor. Recover to level flight as rapidly as possible.

Performance

- Maintain Nz(+) limit for at least 2s. (DESIRED), 1s. (ADEQUATE)
- Transition from Nz(+) pullup to Nz(-) limit pushover within 2s. (DESIRED), 4s. (ADEQUATE)
- Maintain Nz(-) limit for at least 2s. (DESIRED), 1s. (ADEQUATE)

MTE applied modifications

- Maneuver performed for airspeed sweep 140÷220kts
- Target same ΔNz wrt 1g for both pup/pover, equal to ΔNz(+) limit from performance evaluation.
- Time to complete fixed to 6s.: maintain 2s. full long. command pup + 2s. full long. pover + 2s. initial trim long.

Stable short period, ctrl loop not applied.

Piloting logic

- No iterative pilot command computing needed.
- Applied step at each command variation.

Bare A/C approx. 3DoF dynamics [4]

$$\frac{n_z}{\delta_{lon}}(s) = \frac{M_{/\delta lon} Z_{/\alpha}}{\left(s^2 + 2\zeta_{sp}\omega_{sp}s + \omega_{sp}{}^2\right)}$$

$$\frac{q}{\delta_{lon}}(s) = \frac{M_{/\delta lon}\left(s + {}^1/_{T_{\theta 2}}\right)}{\left(s^2 + 2\zeta_{sp}\omega_{sp}s + \omega_{sp}{}^2\right)} \rightarrow \frac{\theta}{\delta_{lon}}(s) = \frac{1}{s}\frac{q}{\delta_{lon}}(s)$$

$$\frac{\gamma}{\theta}(s) = \frac{\left({}^1/_{T_{\theta 2}}\right)}{\left(s + {}^1/_{T_{\theta 2}}\right)} \rightarrow \frac{\gamma}{\delta_{lon}}(s) = \frac{\theta}{\delta_{lon}}(s)\frac{\gamma}{\theta}(s)$$

Definitions:

Flight path-attitude lag: $T_{\theta 2} = {}^V/_{gZ_{/\alpha}}$

$$Z_{/\alpha\_TILT} = \frac{-Z_{/w} massV}{0.5\rho_0 V_{EAS}^2 S_{wing}}$$

$$Z_{/\alpha\_COAX} = \frac{-Z_{/w} mass(V + \Omega_{MR}R_{MR})}{\sigma(\pi R_{MR}^2)(\Omega_{MR}R_{MR})^2}$$

$$Z_{/\alpha\_COWP} = \frac{-Z_{/w\_wing} massV}{0.5\rho_0 V_{EAS}^2 S_{wing}} + \frac{-Z_{/w\_MR} mass(V + \Omega_{MR}R_{MR})}{\sigma\rho_0(\pi R_{MR}^2)(\Omega_{MR}R_{MR})^2}$$

[4] Berger, T., Tischler, M.B., Horn, J.F., "High-Speed Rotorcraft Pitch Axis Response Type Investigation" 77th VFS Annual Forum Proceedings, 2021

Fast Rotorcraft Comparison – High Speed: Pull-up/Push-Over NOE 2/2

TILT: Elevator only
COAX: Lon. Cyclic only
COWP: Lon. Cyclic only

V=150kts

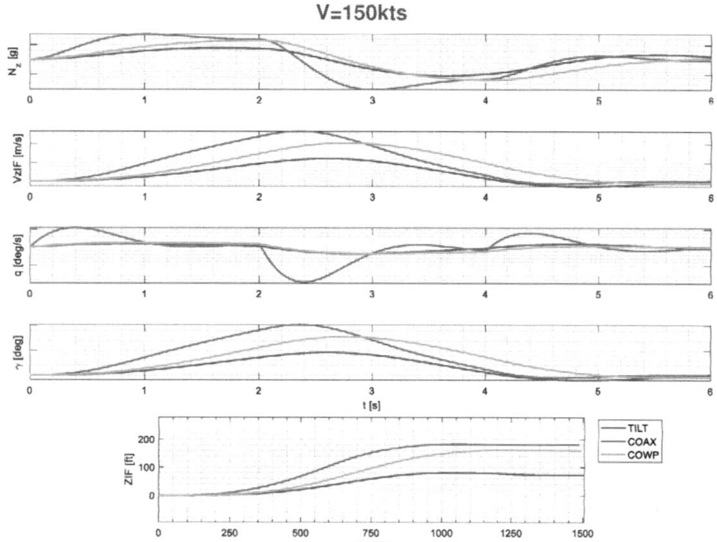

V=200kts

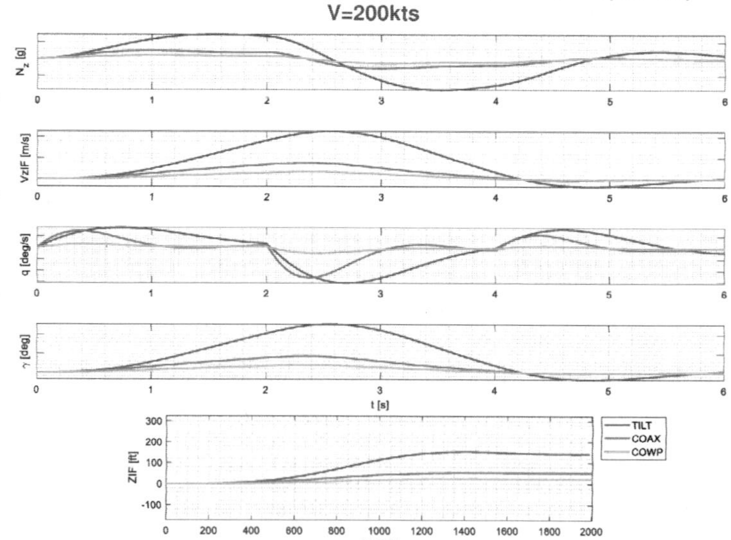

MTEs performance sensitivity analysis to design parameters

Focus on Tiltrotor Architecture

Tiltrotor MTEs performance sensitivity – Lateral Reposition

	MoI I_{xx}	τ_{roll}	$\omega_{dutch-roll}$	$\zeta_{dutch-roll}$	Stab. Deriv. $Y/_v$	Ctrl. Deriv. $L/_{\delta DCP}$ or DCP authority
Design Param. % var.	-30% →+30%	-30% →+30%	-30% →+30%	-30% →+30%	-30% →+30%	-30% → +30%
Time to complete % var.	-4% → +14%	+4% → -2%	-2% → +5%	-0.5% → +2.3%	-1.3% → +0.3%	+23% → -3.5%
Notes	non linear trend, non-symm., higher detrimental effect	non linear trend, non-symm., higher detrimental effect	non linear trend, non-symm., higher detrimental effect	non linear trend, non-symm., higher detrimental effect	non linear trend, non-symm., higher beneficial effect	non linear trend, non-symm., higher detrimental effect

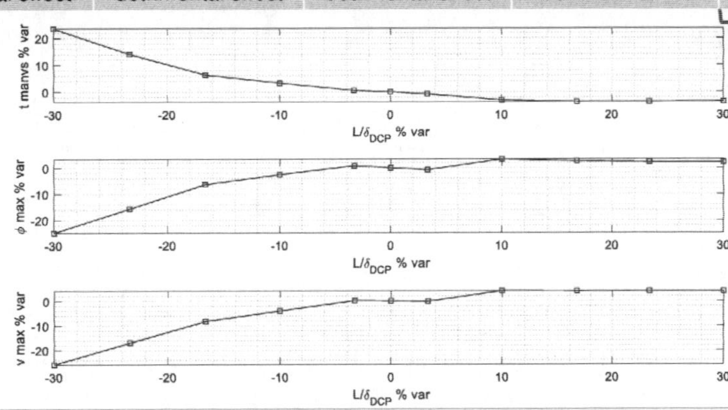

Tiltrotor MTEs performance sensitivity – Pull-up/Push-Over NOE (200kts) 1/2

	Ctrl. Deriv. $M/_{\delta_{elev}}$	Elevator authority	$\zeta_{short\ period}$	Stab. Deriv. $Z/_w$	Stab. Deriv. $M/_w$	Stab. Deriv. $M/_q$	MoI I_{yy}
Design Param. % var.	-30% →+30%	-30% →+30%	-30% →+30%	-30% →+30%	-30% →+30%	-30% →+30%	-30% → +30%
Max height & Avg ramp angle % var.	No change (-17% to +30%) @ -30% = -12%	No change (-10% to +30%) @ -30% = -49%	+1.6% → -1.5%	+0.9% → -1%	+1.2% → -1.4%	+1% → -0.8%	-2.4% → +2.1%
Notes	no change until elevator is saturated (-17% to +30%); change below -17% where Nz target is not achieved	no change until elevator is saturated (-10% to +30%); linear change below -10% where Nz target is not achieved	almost linear & symm trends	almost linear & symm trends before elevator saturation @-30% (discontinuity of -5.5% of max height/avg ramp) ω_{sp} -2% → +2% ζ_{sp} -14% → +14%	almost linear & symm trends ω_{sp} -14% → +12% ζ_{sp} +16% → -11%	almost linear & symm trends ω_{sp} -2% → +2% ζ_{sp} -11% → +11%	almost linear & symm trends ω_{sp} +20% → -12% ζ_{sp} -0.3% → +2%

Negligible sensitivity of max height / avg ramp to stab. deriv, $Z/_q$

Tiltrotor MTEs performance sensitivity – Pull-up/Push-Over NOE (200kts) 2/2

	Target SS ΔN_z max/min
Design Param. % var.	-30% →+30%
Max height & Avg ramp angle % var.	-56.1% → +21.1%

	$\omega_{short\ period}$
Design Param. % var.	-30% →+30%
Max height & Avg ramp angle % var.	+6.7% → -32.5%

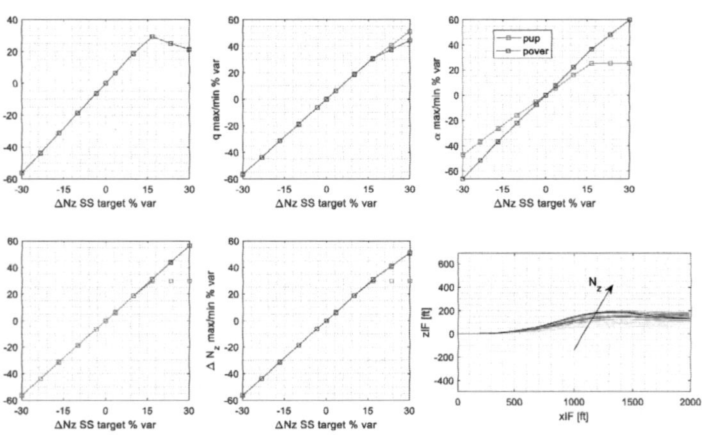

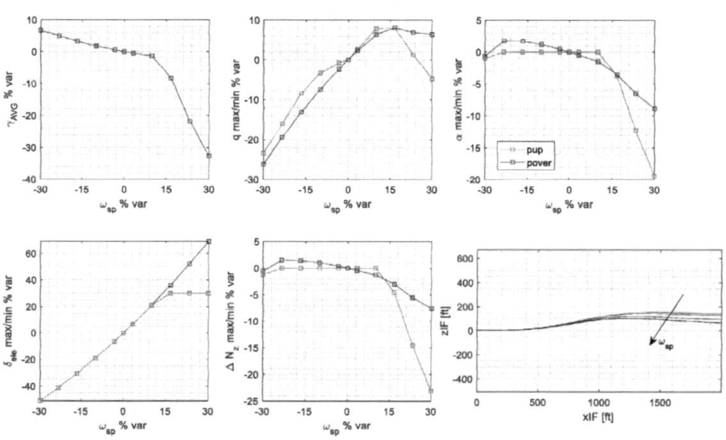

Concluding Remarks & Way Forward

Conclusions

- **A simplified approach is developed to perform preliminary MTEs evaluation in early rotorcraft design**
  - Based on reduced number of bare A/C stability and control derivatives from LTI models (fidelity compatible with design phase)
  - Including a preliminary control loop closure and proper piloting logic
  - Applied as test case on 3 generic fast rotorcraft configurations

- **Goals achieved**
  - Conducted a first guess comparison of different fast rotorcraft configurations performing the same MTE
  - Highlighted design peculiarities of each architecture related to the specific maneuver
  - Evaluated design feedback to improve maneuverability targets at early design phases, through sensitivity analysis on specific parameters mostly impacting the considered MTEs execution performance
  - Identified Control Strategy to improve Fast Rotorcraft control response

Way forward

- Add further MTEs preliminary evaluation, specifically new proposals for low-level & high-speed military flight profiles
- Preliminary Control Loop Design Improvements
- Including Possible Flapping limitation for evaluating MTE
- Enhanced Control Effectors Blending and Control Strategy Implementation
- Execute MTEs with non-linear FLIGHTLAB model, by means of desktop simulations, adopting a similar piloting logic of the developed methodology
- Execute MTEs piloted sessions at simulator facility, running non-linear FLIGHTLAB model with representative FCS

The methodology is intended to contribute at a proprietary multi disciplinary design optimization tool to incorporate a selection of key MTEs performance in rotorcraft preliminary design phase.

LEONARDO

THANK **YOU**
FOR YOUR ATTENTION

leonardo.com

CONTACTS

Leonardo Helicopters

Fabio Riccardi
Senior Flight Dynamics & Control Specialist
fabio.riccardi@leonardo.com

Matteo Pecoraro
Design & Sizing Technical Leader
matteo.pecoraro@leonardo.com

Matteo Mamino
Flight Mechanics Specialist
matteo.mamino@leonardo.com

Andrea Ragazzi
Handling Qualities & Flight Simulation Technical Leader
andrea.ragazzi@leonardo.com

leonardo.com

U.S. ARMY

DEVCOM AVIATION & MISSILE CENTER

PIO AND HANDLING QUALITIES PREDICTION USING THE USAFTPS BJORKMAN PIO DATA SET

Presented at; 2025 VFS Handling Qualities Technical Meeting | 23JAN2025

ED BACHELDER, SENIOR RESEARCH ENGINEER

Distribution A: Approved for Public Release

DEVCOM AVIATION & MISSILE CENTER

PIO and Handling Qualities Prediction Using the USAFTPS Bjorkman PIO Data Set

Ed Bachelder
Senior Research Engineer
San Jose State University
Moffett Field, CA, USA

Eileen Bjorkman
Executive Director (Ret)
Air Force Test Center
Edwards AFB, CA, USA

Bimal Aponso
Chief Scientific & Technical Advisor
Federal Aviation Administration
Seattle, WA, USA

Distribution A: Approved for Public Release

OUTLINE

DEVCOM AVIATION & MISSILE CENTER

- Objectives
- Technology Capability Gaps
- Bjorkman & VMS PIO Studies
- HQ Metric: Decay Rate Parameter
- Pilot Model
- PIO Frequency Prediction Results
- Decay Rate Threshold
- Rating Prediction Results
- Contribution to Research
- Operational Impact
- Future Work

RESEARCH OBJECTIVES

- **Develop a quantitative rationale to predict PIO and HQ rating variability**
  - Understand fundamental principles governing pilot rating
- **Develop a model to predict inflight and simulation pilot ratings**

Configu-ration	# of Flights	PIO Ratings	HQ Ratings
2-B	4	3/2/2/1	7/3/3/3
2-1	3	1/1/1	2/2/3
2-5	3	4/4/5	10/7/10
2-7	3	4/3/2	7/4/4
2-8	3	4/4/4	8/10/8
3-D	2	1/1	2/2
3-1	3	3/2/2	5/3/4
3-3	3	3/1/1	7/2/3
3-6	2	2/2	5/4
3-8	3	4/3/4	8/5/8
3-12	2	4/5	7/9
3-13	2	4/5	10/10
4-1	3	1/1/1	3/2/3
4-2	3	1/1/2	3/3/7
5-1	2	1/1	2/5
5-9	2	4/4	7/7
5-10	2	5/5	10/10
5-11	3	2/4/3	7/7/5

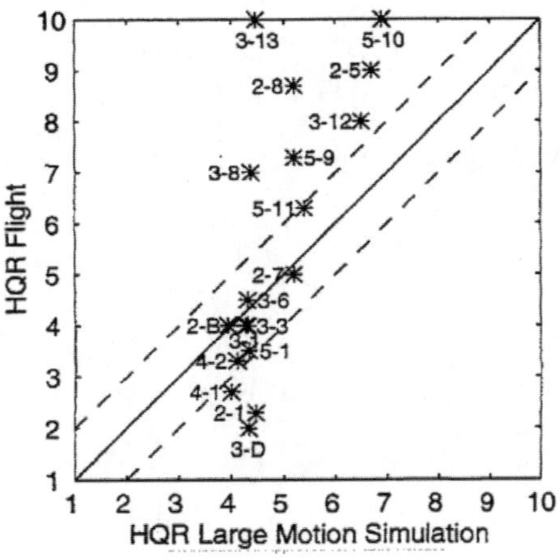

TECHNOLOGY CAPABILITY GAPS

- **A typical trigger event for pilot induced oscillation (PIO) is the passing of control from automation to the pilot, which often happens when the automation fails**
- **Although PIOs were observed as early as the Wright brothers, its mechanisms are still not well-understood**
- **No deterministic model accounting for variability in pilot ratings**
- **Cooper-Harper is a multi-dimensional rating scale that employs difficult-to-quantify terms such as workload, compensation, and controllability**
- **No general theory or model predicts differences in control between inflight and simulated operation**

BJORKMAN PIO STUDY (1986)

- **USAF/Calspan variable stability NT-33A aircraft**
- **18 aircraft configurations using Flight Control System lag filters**
- **Offset landing approach task**
- **Three test pilots, 126 approaches**

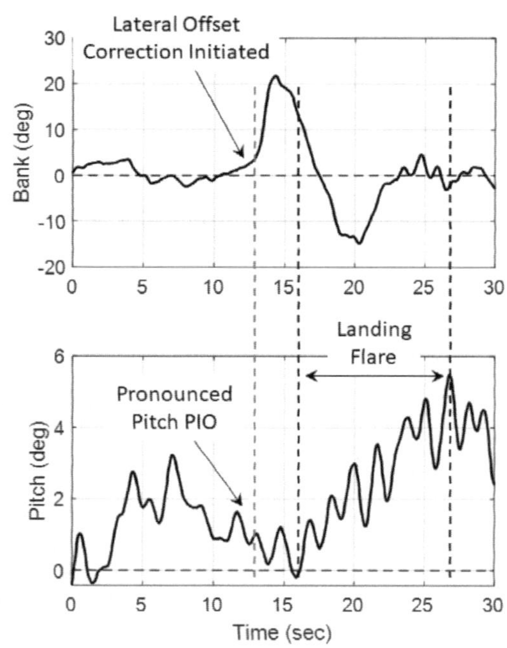

NASA VMS MOTION STUDY USING BJORKMAN CONFIGURATIONS (1998)

DEVCOM AVIATION & MISSILE CENTER

- **Replicated Bjorkman study's landing task and eighteen aircraft configurations**
- **NASA Vertical Motion Simulator (VMS) has the largest motion travel envelope of any ground-based flight simulator**
  - Six independent degrees-of-freedom
- **Three levels of simulation motion platform displacement**
  - Large
  - Small
  - None (visual)
- **5 test pilots, over 1,400 approaches**

8

PIO RATING SCALE

Description	Rating
No tendency for pilot to induce undesirable motions	1
Undesirable motions tend to occur when pilot initiates abrupt maneuvers or attempts tight control. These motions can be prevented or eliminated by pilot technique.	2
Undesirable motions easily induces when pilot initiates abrupt maneuvers or attempts tight control. These motions can be prevented or eliminated but only at sacrifice to task performance or through considerable pilot attention and effort.	3
Oscillations tend to develop when pilot initiates abrupt maneuvers or attempts tight control. Pilot must reduce gain or abandon task to recover.	4
Divergent oscillations tend to develop when pilot initiates abrupt maneuvers or attempts tight control. Pilot must open loop by releasing or freezing stick.	5
Disturbance or normal pilot control may cause divergent oscillations. Pilot must open control loop by releasing or freezing stick.	6

Simulate abrupt maneuver using a step pitch command

HQ HYPOTHESIS

Following abrupt control motion, the dominant influence on performance is the length of time the oscillation remains large enough to interfere with the task

Damped oscillation bounded over time by the envelope $\pm e^{-\alpha t}$

α : exponential decay rate $= \zeta\omega_n$
ζ : damping ratio
ω_n : natural frequency

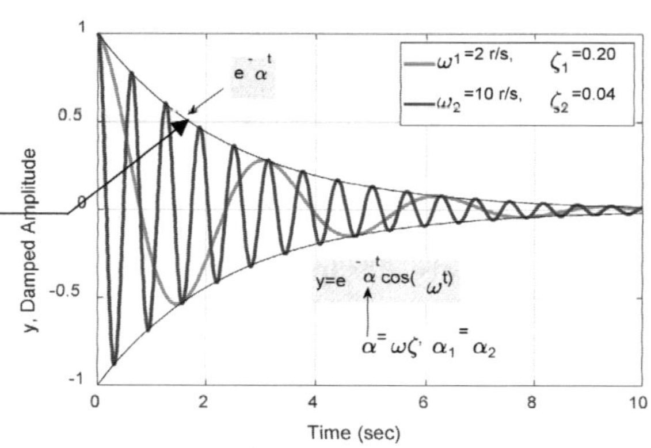

PIO ratings are the same for both oscillations

COMPENSATION AND REGULATION PILOT MODEL

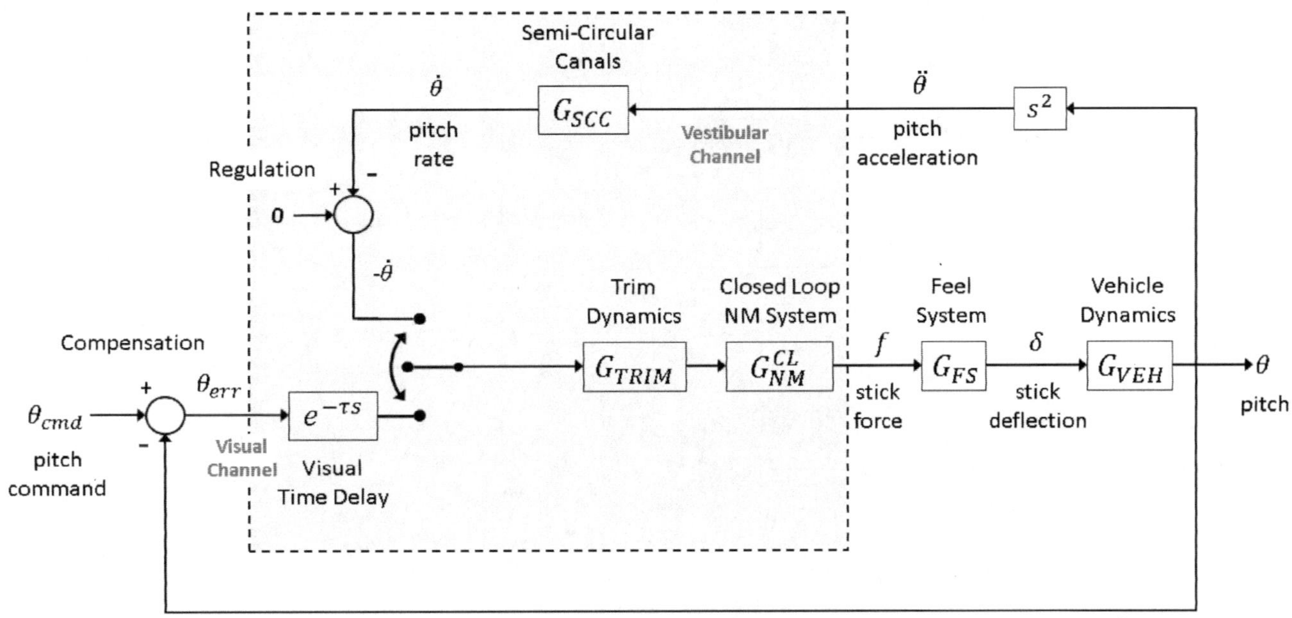

SIMULATION TECHNIQUE

- **Start with the known vehicle dynamics and initialized pilot model**
- **Apply unit step command to closed loop (CL) visual channel (switch is in down position)**
- **Iterate on pilot model parameters to minimize difference between pitch step command and pitch output**
- **Obtain decay rate of CL rigid body mode**
- **Repeat for vestibular channel**

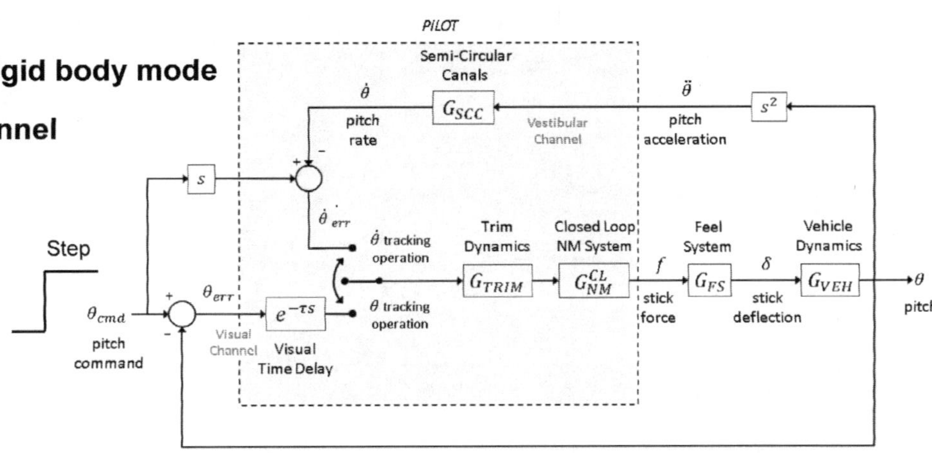

PIO FREQUENCY ANALYSIS

Flight Data Histories

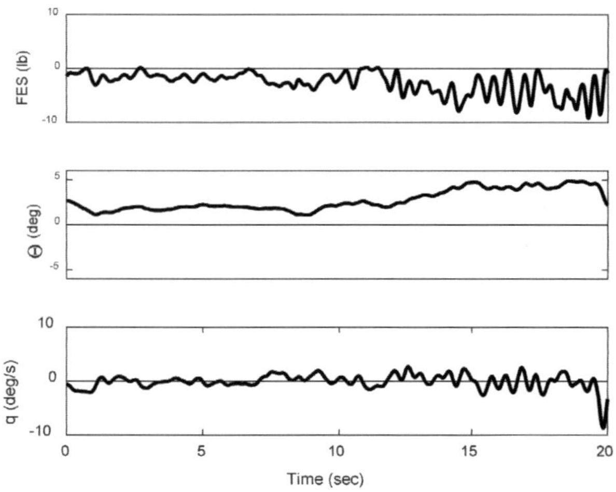

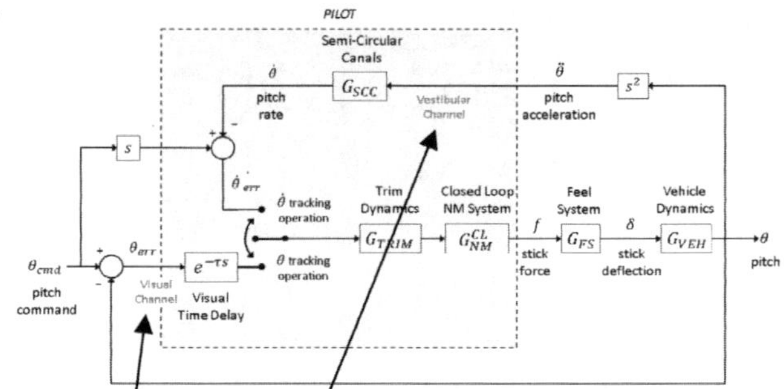

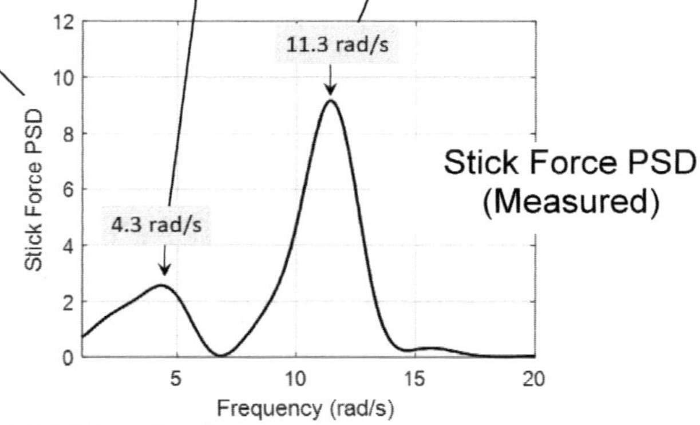

Lower frequency associated with θ tracking (visual channel)

Higher frequency associated with $\dot{\theta}$ tracking (vestibular channel)

VESTIBULAR CHANNEL MODELING

Closed loop response (red) using $\dot{\theta}$ tracking channel of pilot model

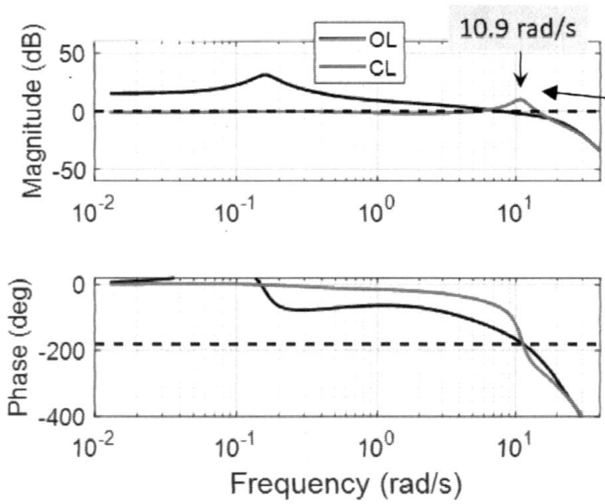

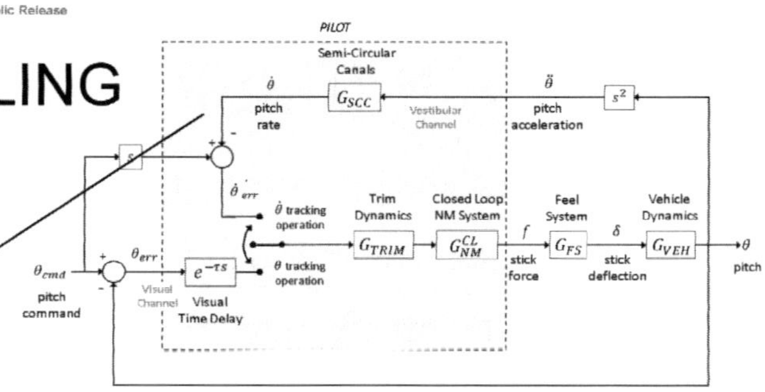

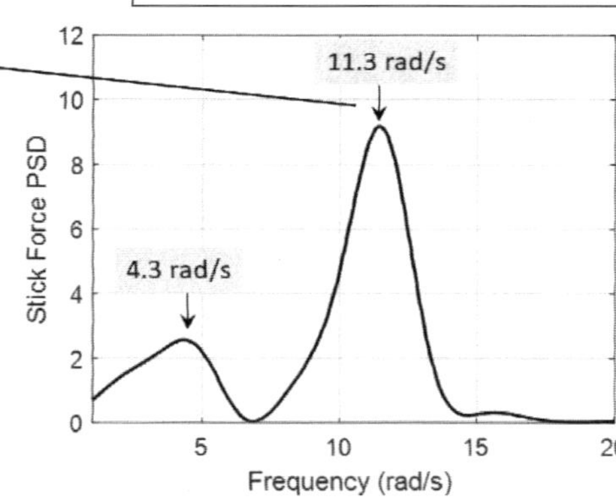

Stick Force PSD (Measured)

VISUAL CHANNEL MODELING

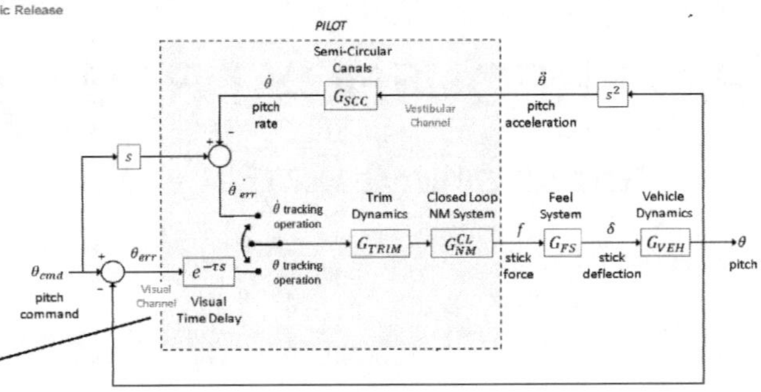

Closed loop response (red) using
θ tracking channel of pilot model

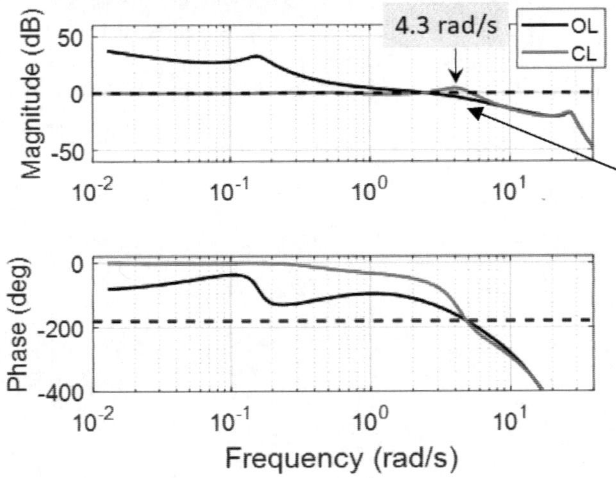

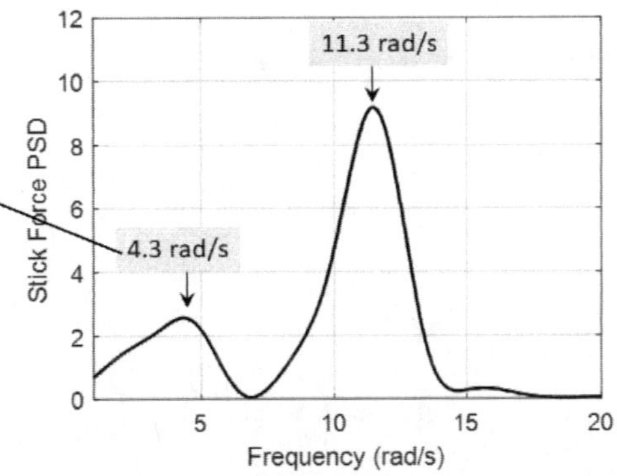

Stick Force PSD
(Measured)

MATCHING VERSUS CORRELATION COEFFICIENT

- The Pearson Correlation Coefficient *R* measure is a metric characterizing how well data matches a best-fit line through the data
- With prediction, the interest is to measure how well a prediction matches the observation

Matching coefficient

$$R_{match} = \sqrt{1 - \sqrt{\frac{\sum (Y_{obs} - Y_{pred})^2}{\sum Y_{obs}^2}}}$$

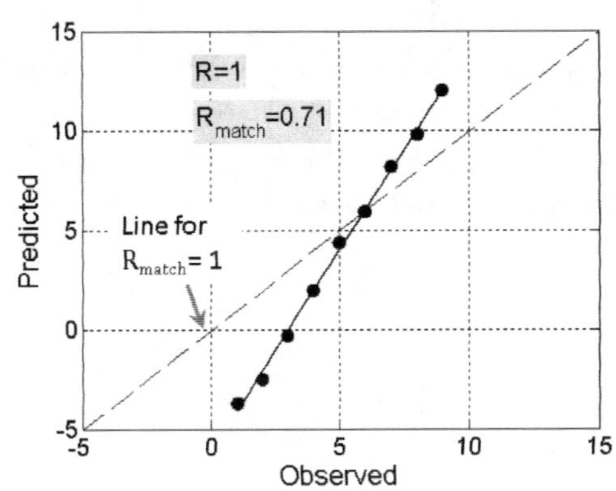

Distribution A: Approved for Public Release

PIO FREQUENCY PREDICTION

DEVCOM AVIATION & MISSILE CENTER

Two peaks observed during PIO

Higher freq mode linked to $\dot{\theta}$ tracking

Lower freq mode linked to θ tracking

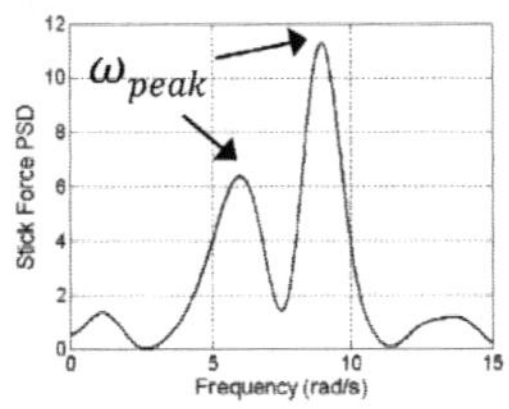

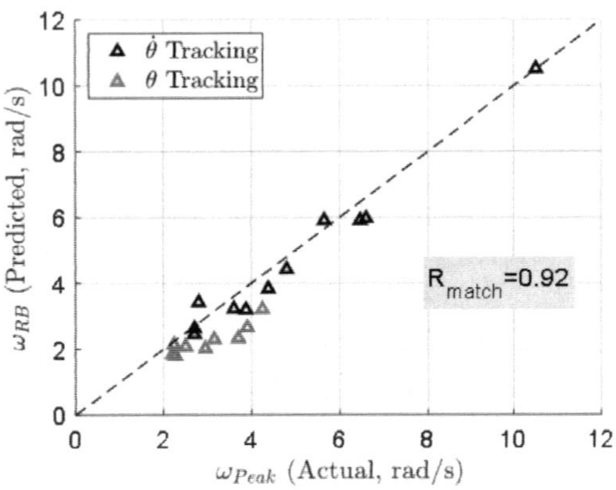

Single peak observed during PIO

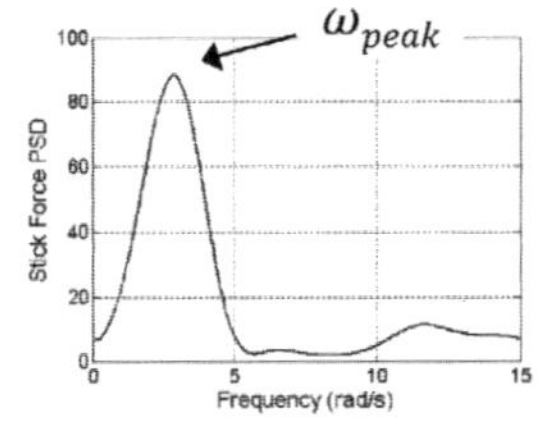

Mode linked to tracking strategy that is closest to it

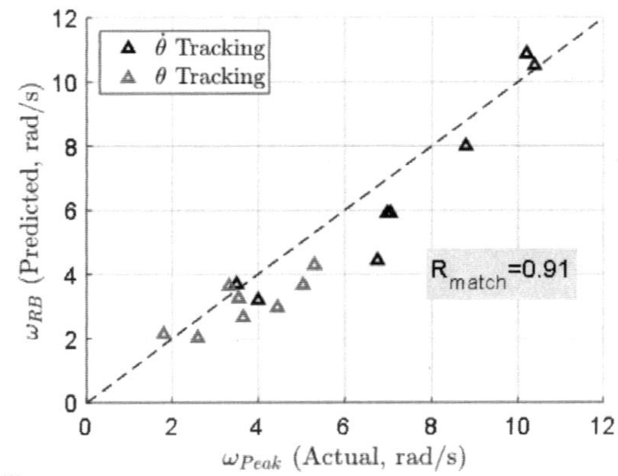

Distribution A: Approved for Public Release

17

Distribution A: Approved for Public Release

TRACKING ERROR VS. DECAY RATE THRESHOLD

DEVCOM AVIATION & MISSILE CENTER

- **Point of maximum curvature between RMS error and α is hypothesized to be where the pilot first perceives a change in task performance**
  - Analogous to the just noticeable difference (JND) in psychophysics

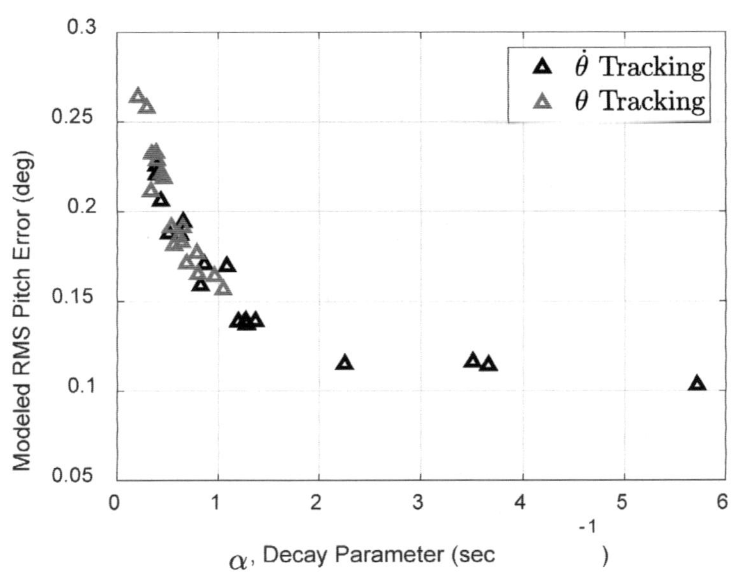

Distribution A: Approved for Public Release

18

HYPERBOLA FITTED TO RE-SCALED DATA

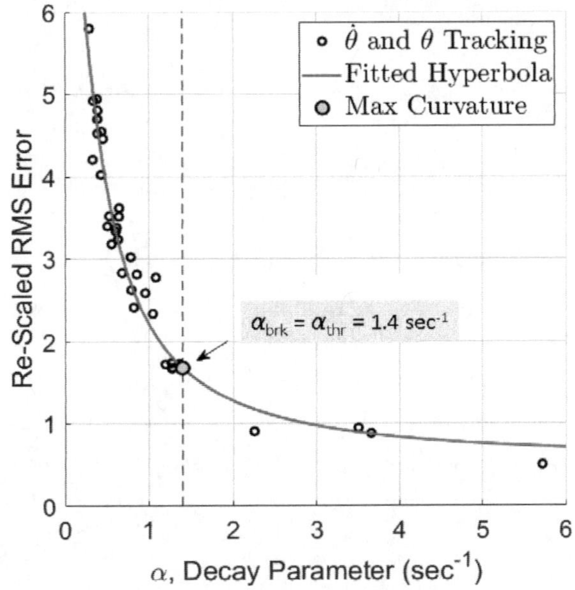

Break point at $\alpha_{\text{brk}} = 1.4$ denoted as decay rate threshold (α_{thr})

19

PIOR PREDICTION TEMPLATE CONSTRUCTION

6	Disturbance or normal pilot control may cause divergent oscillations. Pilot must open control loop by releasing or freezing stick.

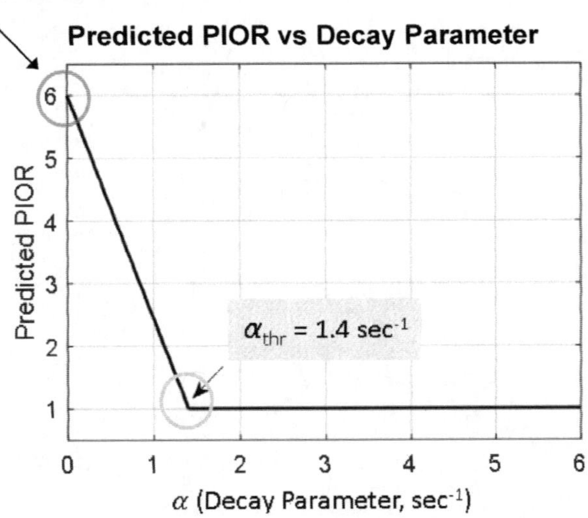

FITTING ACTUAL PIOR TO DECAY PARAMETER

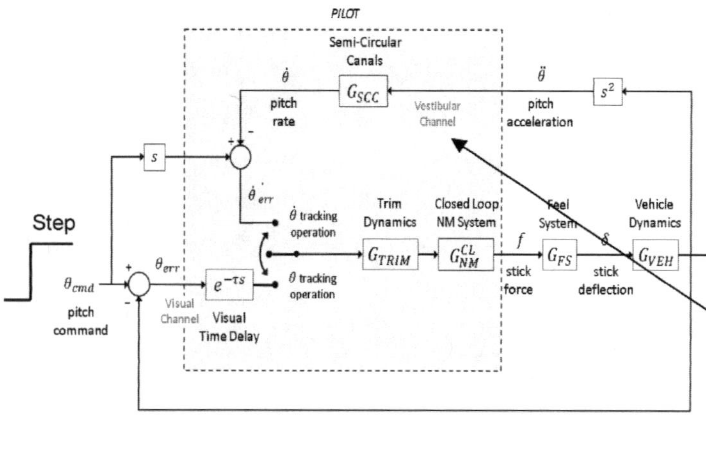

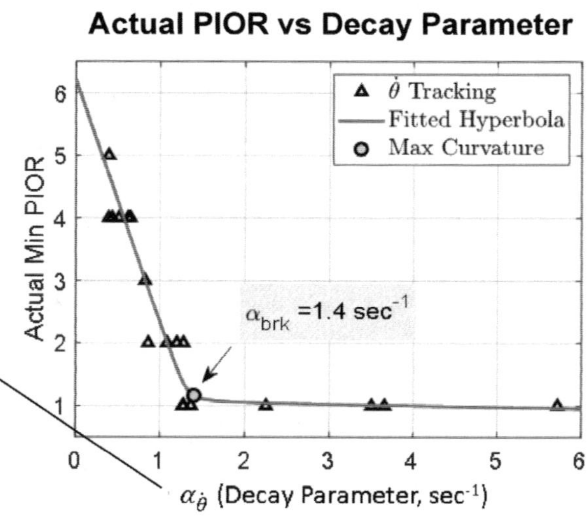

PIOR PREDICTION TEMPLATE COMPARISON

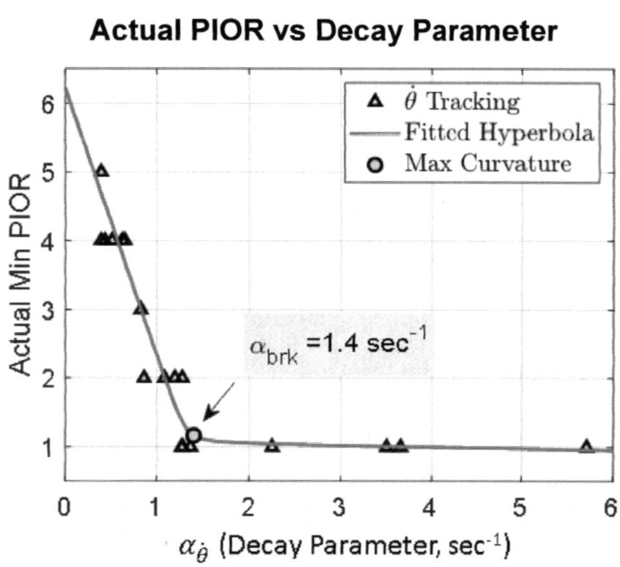

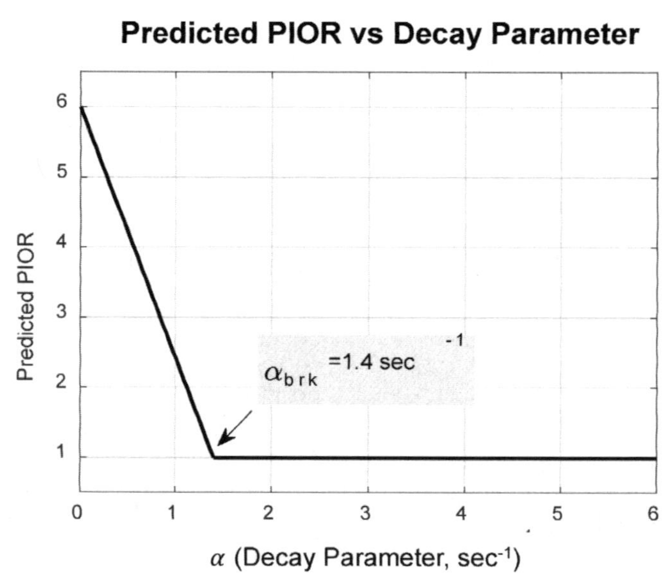

PIO RATINGS PREDICTION

Minimum PIOR Uses $\alpha_{\dot{\theta}}$

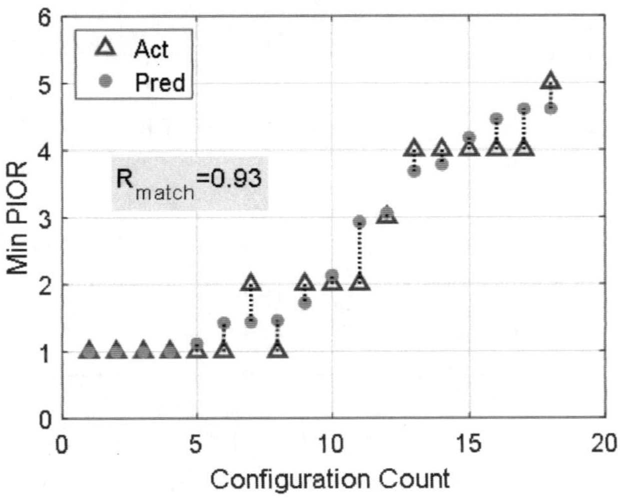

Maximum PIOR Uses α_{θ} and $\alpha_{\dot{\theta}}$

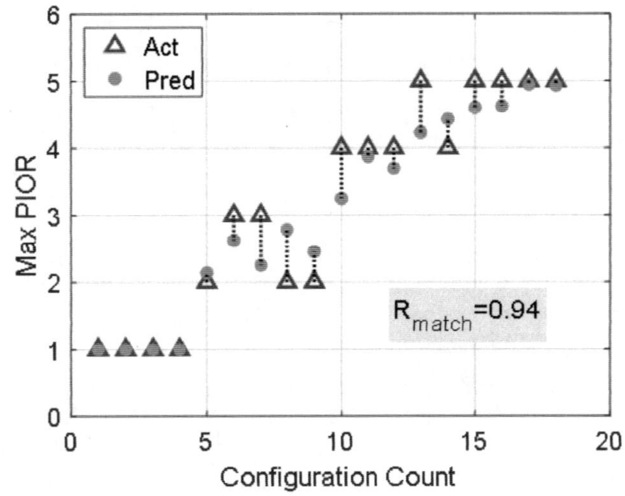

Distribution A: Approved for Public Release

Distribution A: Approved for Public Release

HQR PREDICTION TEMPLATE CONSTRUCTION: *PIOR5 LINKED TO HQR10*

PIO Rating Scale

Divergent oscillations tend to develop when pilot initiates abrupt maneuvers or attempts tight control. Pilot must open loop by releasing or freezing stick.	5

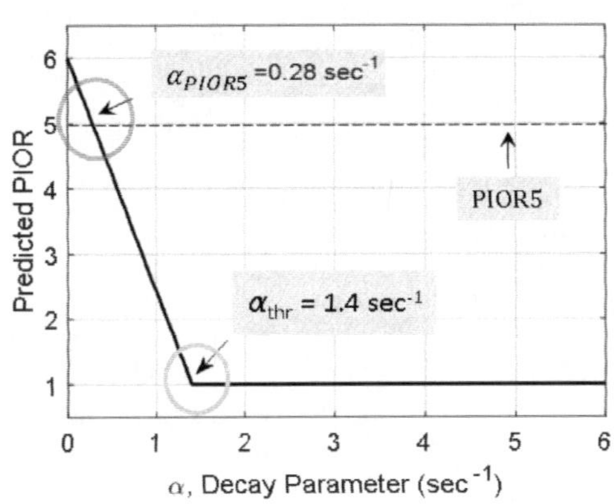

HQ Rating Scale

Control will be lost during some portion of the required operation	10

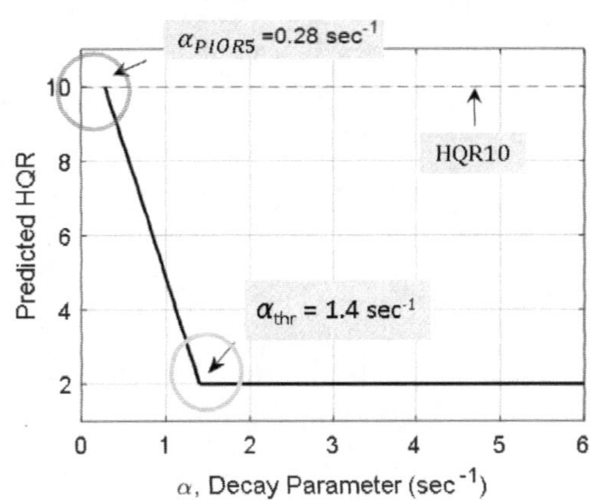

Distribution A: Approved for Public Release

HQR PREDICTION TEMPLATE COMPARISON

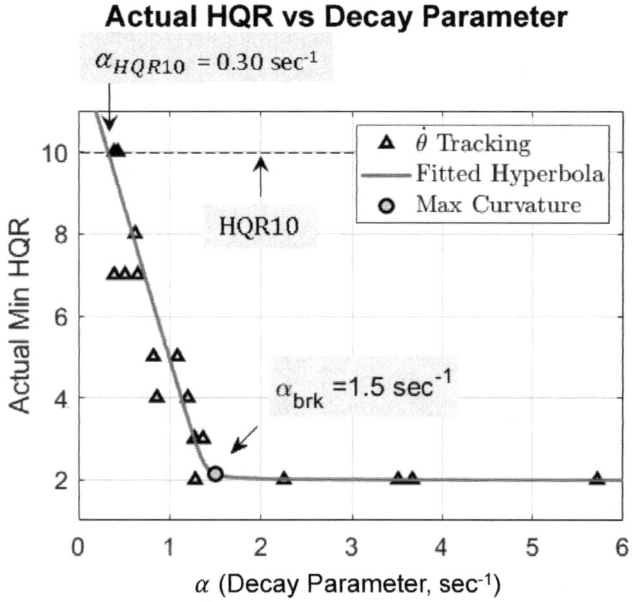

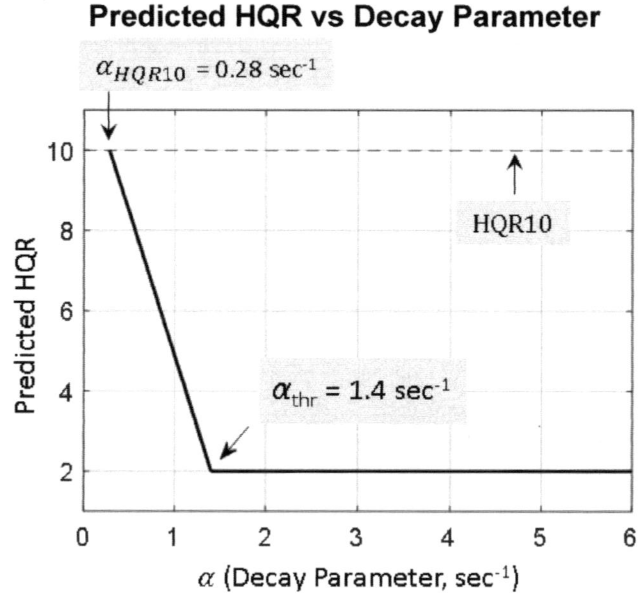

HQ RATINGS PREDICTION

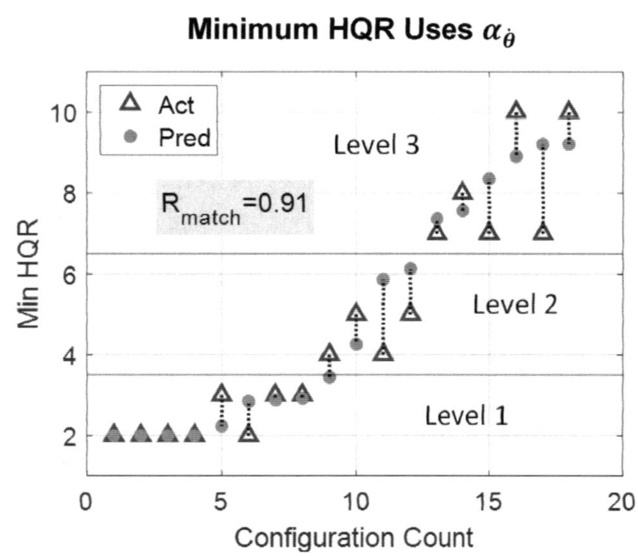

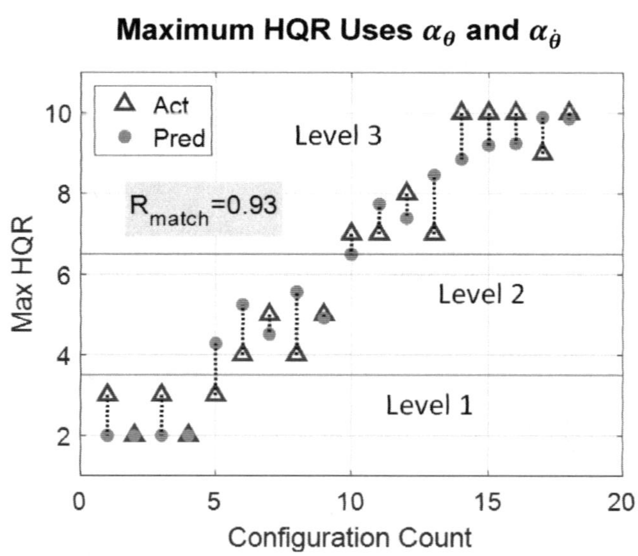

INFLIGHT VS VMS (LARGE MOTION)

PIOR

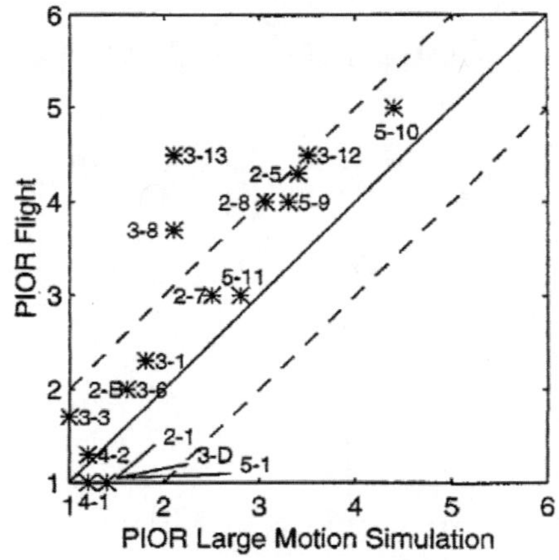

$PIOR_{VMS} < PIOR_{Flight}$

Significant PIOR bias

HQR

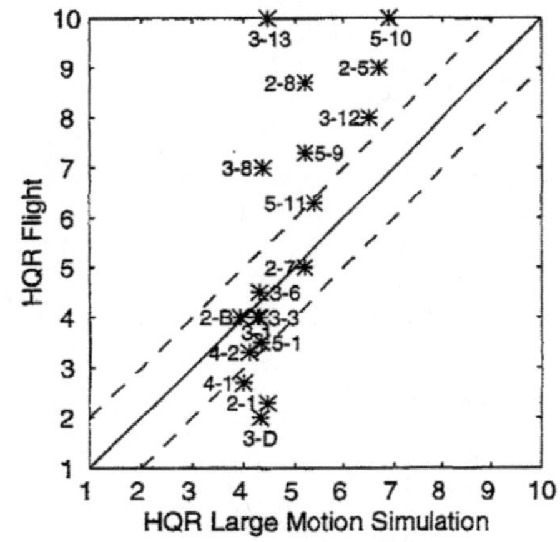

Min/Max HQR_{VMS} = 4/7

Min/Max HQR_{Flight} = 2/10

Significant HQR skewing

VMS LARGE MOTION VS FIXED-BASE RATINGS

PIOR

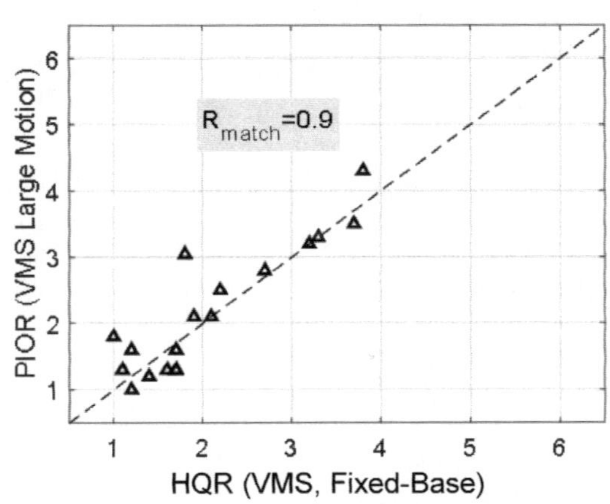

HQR

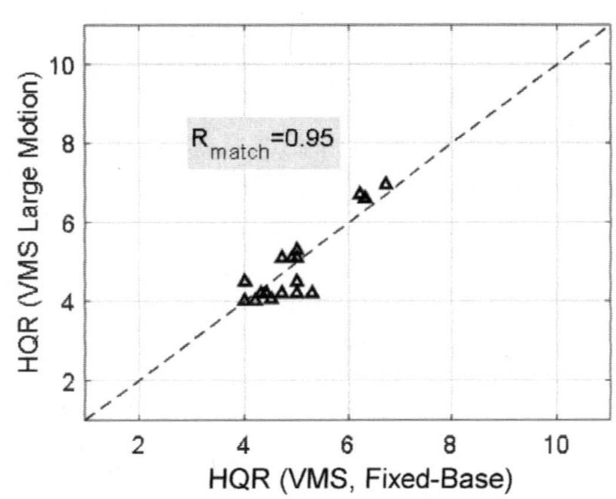

Large motion and fixed-base ratings very similar

FITTING VMS PIOR TO DECAY PARAMETER

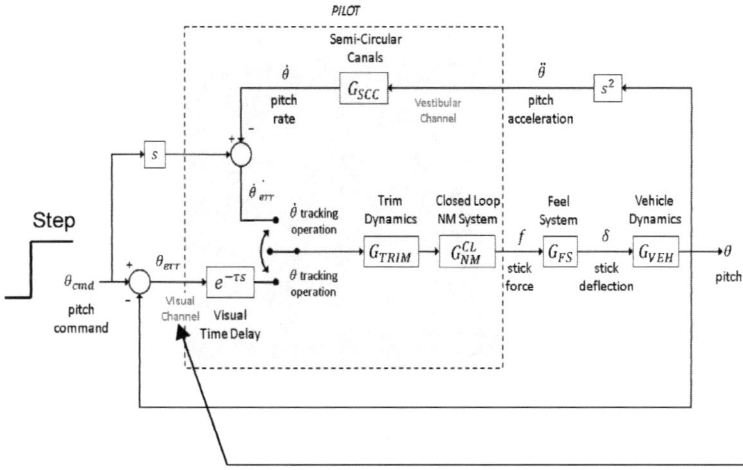

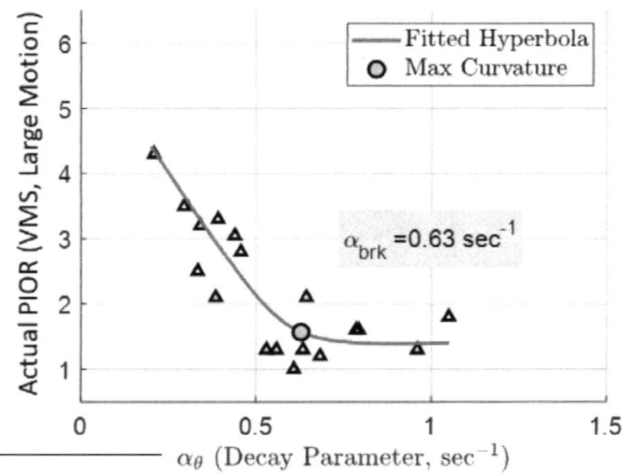

VMS PIOR break at α = 0.63 sec^{-1} (1.4 sec^{-1} for flight)

VMS AND FLIGHT PIOR TEMPLATES

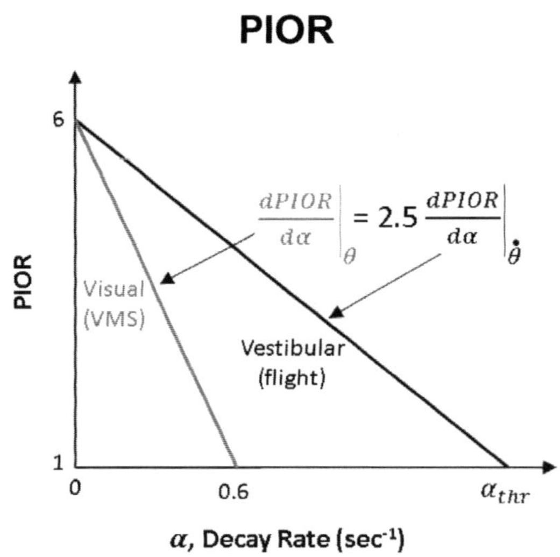

Expect higher prediction uncertainty with VMS PIO ratings than flight

PIO RATING PREDICTION (VMS & FLIGHT)

DEVCOM AVIATION & MISSILE CENTER

VMS PIOR

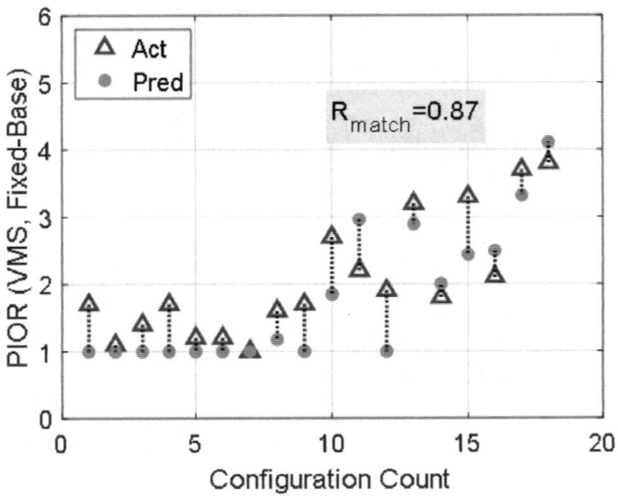

Flight PIOR

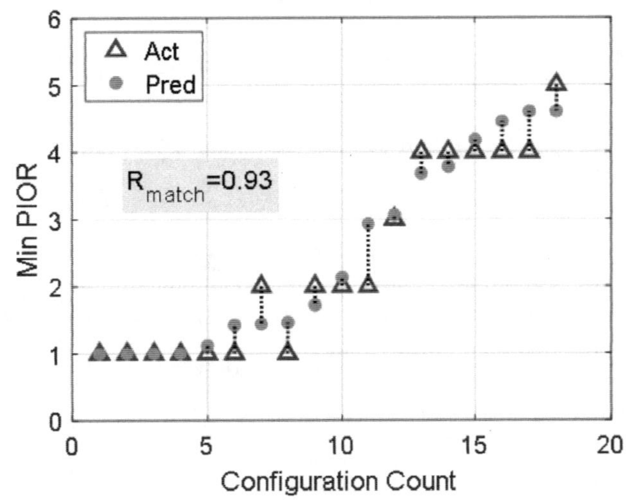

VMS AND FLIGHT HQR TEMPLATES

DEVCOM AVIATION & MISSILE CENTER

HQR

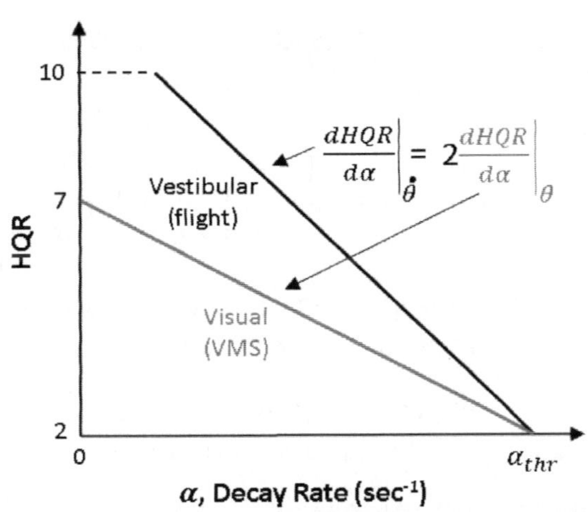

Expect less prediction uncertainty with VMS HQ ratings than flight

HQ RATINGS PREDICTION (VMS & FLIGHT)

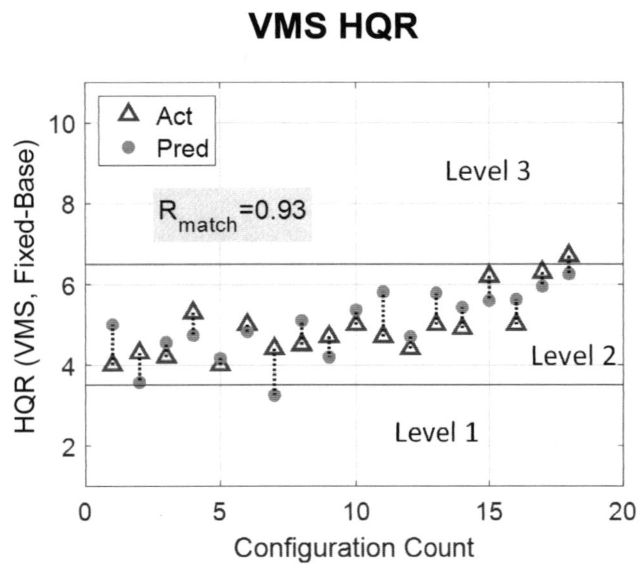

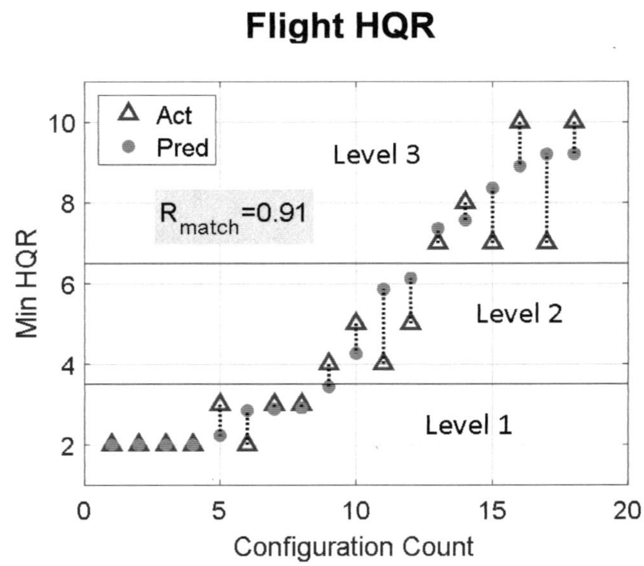

CONTRIBUTION TO RESEARCH

- **Decay rate of the closed loop rigid body mode is proposed as a handling qualities metric**
- **Vestibular and visual channels of the pilot model produced decay rates that closely match best and worst inflight ratings**
  - PIO frequency prediction was excellent
- **Dominant channel appears to be vestibular (i.e., pitch rate feedback used) for a precision flight task that is anchored in the physical, out-the-window environment**
- **The pilot relies primarily on the visual channel (i.e., pitch rate feedback NOT used) during flight simulation irrespective of simulator motion fidelity**
- **Inflight tasks using electronic displays may have limitations akin to those of simulators due to mismatch between the lagged visual and real time vestibular cueing**
- **Decay rate parameter is relevant for automated systems**

OPERATIONAL IMPACT

- Reduce costs associated with schedule delays
- Smaller flight test matrix
- Design tool for maximizing pilot useability of a flight system's features
- Optimize design to minimize weight
- Optimize allocation of simulation and flight resources for training
- Generate avoid regions of operation in existing and future aircraft

Distribution A: Approved for Public Release

FUTURE WORK

- 2025 VFS paper
  - Decay rate as a HQ and PIO predictor will be further validated using additional Calspan flight data (lateral HQs, roll ratchet, 2019 Learjet study)
  - Inflight tasks using electronic displays will be shown to have limitations akin to simulation (only visual channel used)
  - Establish how ADS-33 phase delay (PD) and bandwidth (BW) parameters are related to decay rate, and derive the limitations of PD and BW
- Set up for inflight validation of pilot model and the role of decay rate in HQs using RASCAL-X
- PIO paper and 72 other publications are available on Research Gate

QUESTIONS

Distribution A: Approved for Public Release

Distribution A: Approved for Public Release

BACKUP SLIDES

Distribution A: Approved for Public Release

Distribution A: Approved for Public Release

PIO RATINGS PREDICTION (VMS)

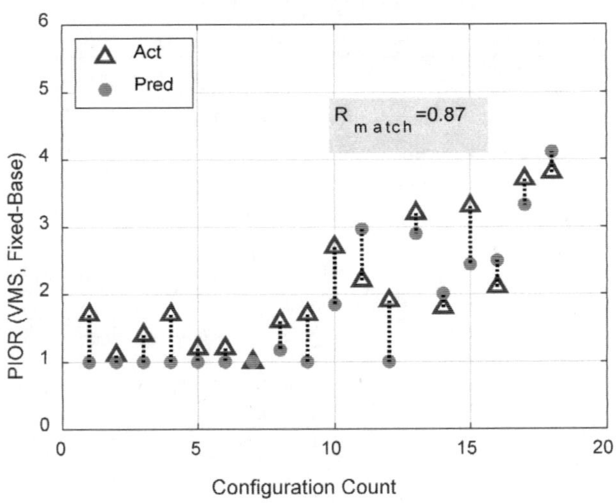

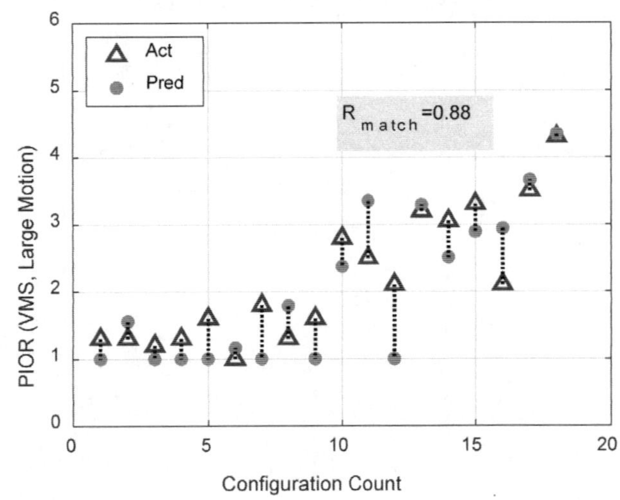

Distribution A: Approved for Public Release

39

Distribution A: Approved for Public Release

HQ RATINGS PREDICTION (VMS)

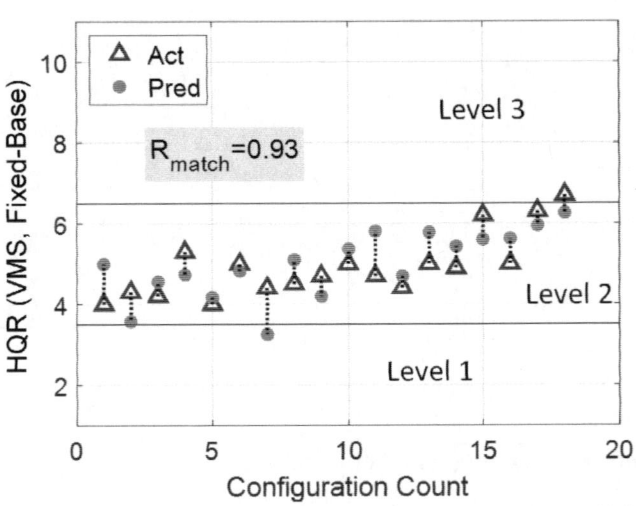

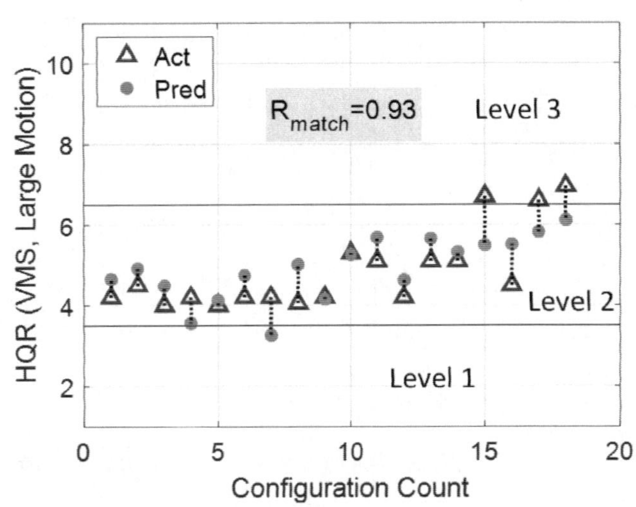

Distribution A: Approved for Public Release

40

EFFECTIVE DECAY RATE

DEVCOM AVIATION & MISSILE CENTER

Configuration Decay Rate

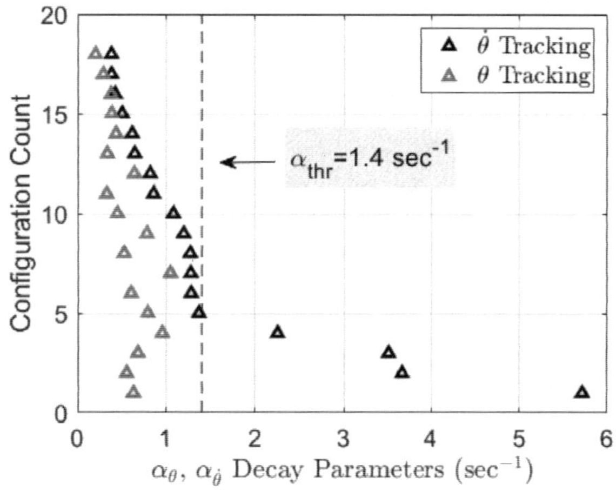

Effective Decay Rate

Min PIOR. HQR: $\alpha_{eff} = \alpha_{\theta}$

Max PIOR, HQR:

$$\begin{cases} \alpha_{\dot{\theta}} > \alpha_{thr} & \alpha_{eff} = \alpha_{\dot{\theta}} \\ \alpha_{\dot{\theta}} < \alpha_{thr} \;\&\; \dfrac{|\alpha_{\dot{\theta}} - \alpha_{\theta}|}{\alpha_{\dot{\theta}}} < 0.3 & \alpha_{eff} = \alpha_{\theta} \\ \alpha_{\dot{\theta}} < \alpha_{thr} \;\&\; \dfrac{|\alpha_{\dot{\theta}} - \alpha_{\theta}|}{\alpha_{\dot{\theta}}} > 0.3 & \alpha_{eff} = \dfrac{(\alpha_{\theta} + \alpha_{\dot{\theta}})}{2} \end{cases}$$

VISUAL AND VESTIBULAR DECAY RATES

DEVCOM AVIATION & MISSILE CENTER

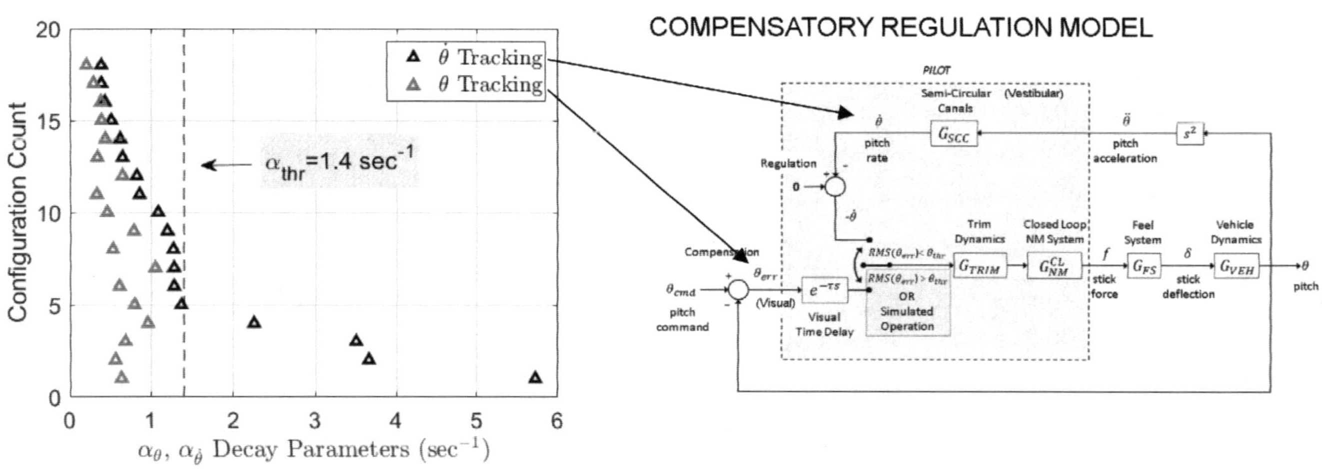

VMS (fixed-base, large motion) linked with θ tracking (visual channel)

Flight linked with $\dot{\theta}$ tracking (vestibular channel)

VMS AND FLIGHT RATING TEMPLATES

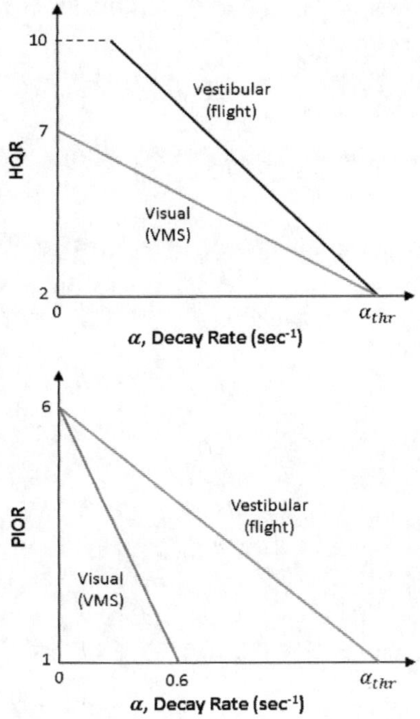

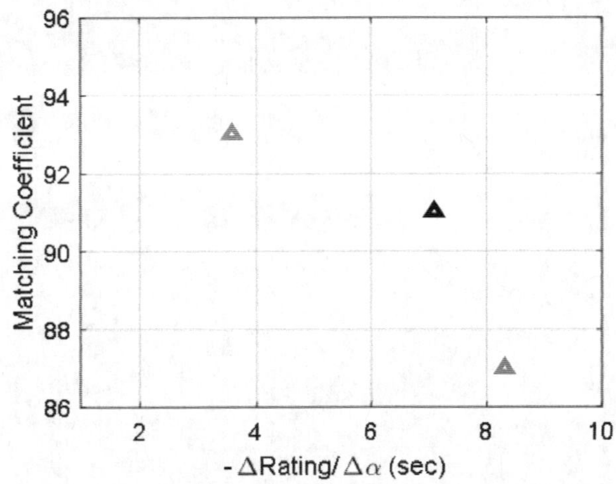

Rating variability linked to rating slope

RATINGS PREDICTION (VMS & FLIGHT)

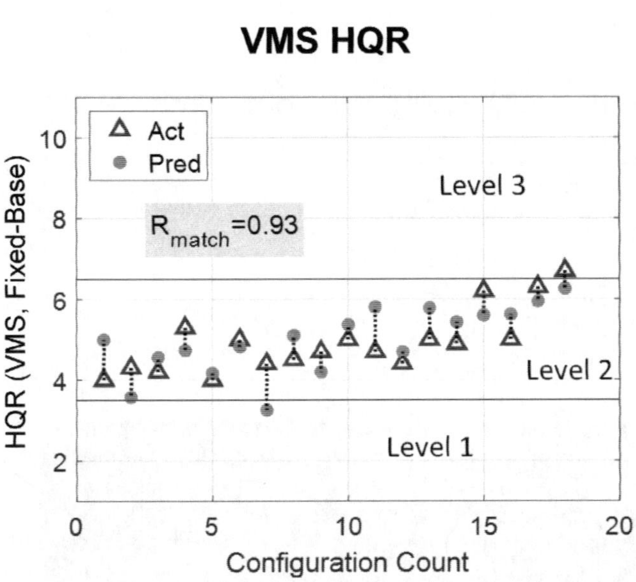

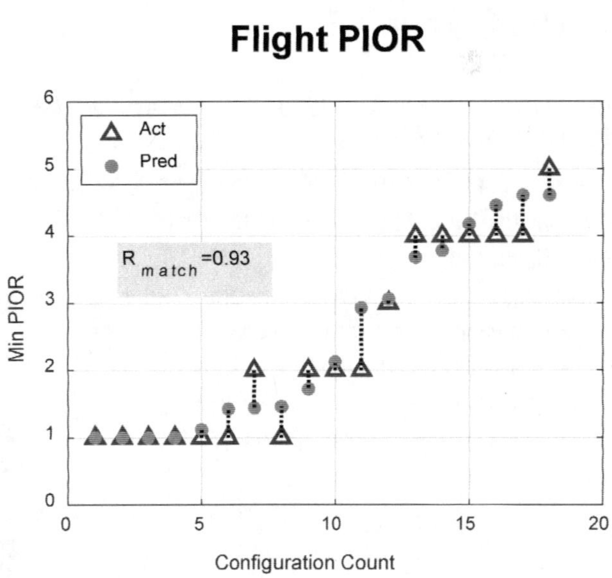

INTUITIVE®

Usability Shortcomings in Department of Defense Funded Systems

Jerry Burpee
Usability Research and Evaluation
Sr. Principal Engineer
January 23, 2025

Why DoD Funded Systems Experience Usability Shortcomings

INTUITIVE®

Usability is lacking in DoD developed systems compared to public or commercial developed systems.

These shortcomings are partially due to the following factors:

- Funding Models are Different
- Limited Competition
- Requirements and Specifications are Hard to Define
- Usability Metrics are Hard to Define
- Specialized Users
- Lack of Usability Specialists
- Non-Revenue Generation

> Usability is the "*extent to which a system, product or service can be used by specified users to achieve specified goals with effectiveness, efficiency, and satisfaction in a specified context of use.*"
>
> ISO 9241-11:2018 Ergonomics of human-system integration – Part 11: Usability: Definitions and concepts, 2018, International organization of Standards

UNCLASSIFIED, DISTRIBUTION A

2

Funding Models are Different

INTUITIVE®

Government Projects are funded <u>then</u> the system is built.

Private Sector Projects are built <u>then</u> purchased.

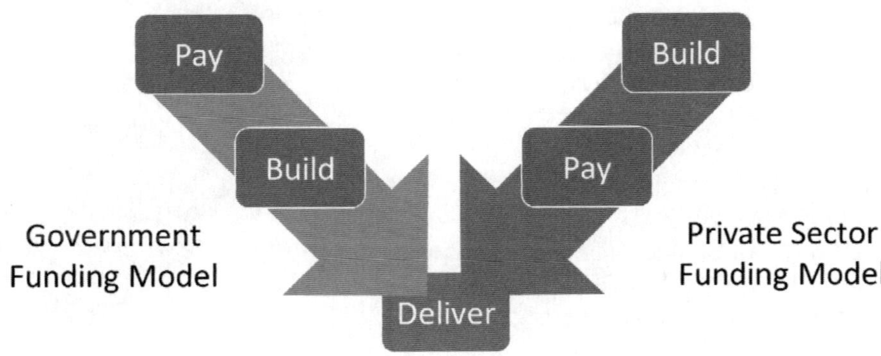

Limited Competition

INTUITIVE®

Competition for developing a government system is concentrated during the bidding process before the project is awarded.

- If usability requirements and specifications are not defined in the bid process, usability issues discovered later are not disqualifiers.

Minimal incentive to discuss usability prior to bid.

- Contract add-on incentives for follow-on work to fix usability issues.

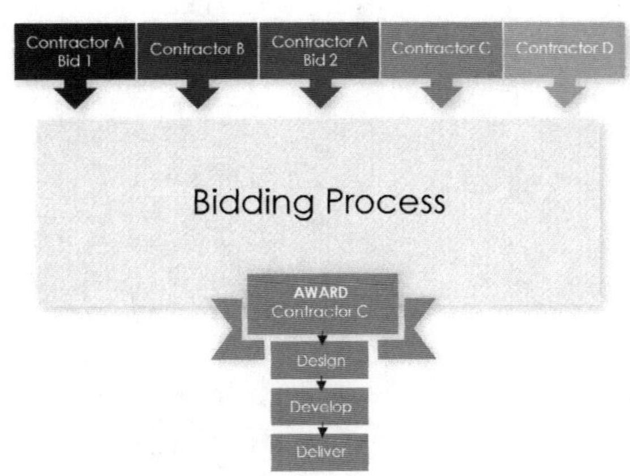

1. Request for bids based on requirements and specifications is solicited.
2. Several entities submit bids.
3. A contract is awarded to a winning bid prior to system design.

Requirements & Specifications are Hard to Define

- Properly written usability requirements and specifications are key factors to reaping the benefits of a user-friendly product.
- Most requirements and specifications are objectively written with defined metrics to be met, proven, or confirmed during later stages of design.
  - o Test results are often simple binary responses such as pass/fail, or results must fall within a defined threshold, such as ±3 degrees, not by measurements of effectiveness and efficiency.
- Defining usability metrics in requirements and specifications can be very difficult.
  - o Variables such as method, environment, samples sets, prototype fidelity need to be defined.
  - o Expectations can become too vague or too restrictive.

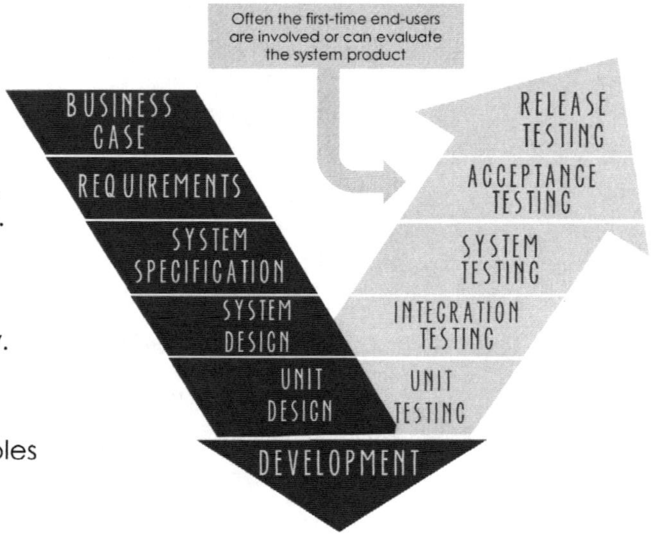

Usability Evaluation versus System Test

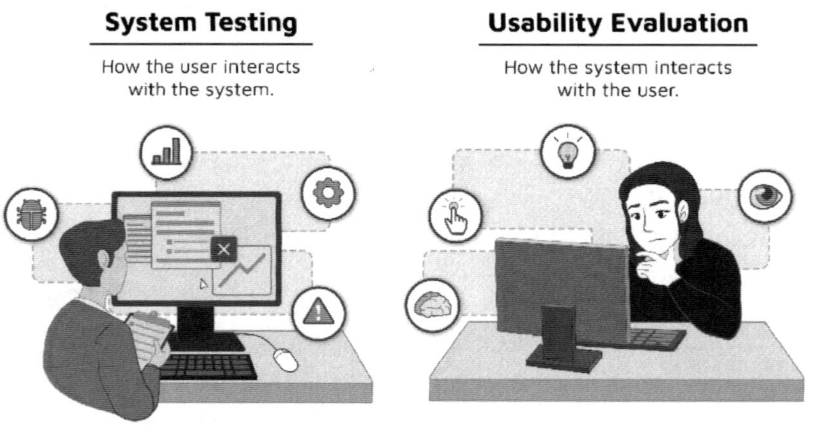

Focus = SYSTEM	Focus = USER
•Does the system meet the requirements?	•Does the system meet the user's needs?
•Can the user perform the necessary steps to make the system work?	•How intuitive is it for the user to perform the necessary steps to make the system work?
•Results are typically "Pass" or "Fail".	•Results are typically measures of effectiveness and efficiency.

Burpee, Tenhundfeld; Usability Evaluations & Usability Laboratories; 2023, Amazon

A Solution: Define the Scope of Work (SoW)

INTUITIVE

- Project shall have a dedicated User Experience (UX) Professional(s) involved "from initial user requirements through the program life cycle to system disposal"*.
- The UX Professional(s) shall define user needs prior to design and development.
- The UX Professional(s) shall be involved in the product research and discovery phase, and provide recommendations to design, development, and training.
- The UX Professional(s) shall conduct "usability and other user testing to support and inform human and machine interface analysis under operational conditions"* prior to or during specified milestones.
- The UX Professional(s) shall provide the design and development team heuristic analysis of user interactions and interfaces.
- The UX Professional(s) shall conduct a benchmark evaluation at stage XX (or XX% product completion) and another one at completion of product development.

NOTE: As of April 1, 2022
DoD Instruction 5000.95
Human System Integration in Defense Acquisition
provides usability requirements that DoD is directed to implement

ALSO: See **DoD HSI Guidebook** (May 2022)
for additional usability implementation guidance

* Quoted from DoDI 5000.95

Usability Metrics are Hard to Define

INTUITIVE

Usability is often interpreted to pertaining to the aesthetics of the interface display. However, usability involves the entire interaction between the system and the user.

Usability Metrics Use Key Performance Indicators (KPIs):

- Task success or completion rate
- Time on task
- User error rate
- Product usefulness
- Product usability
- User workload (physical or cognitive)
- User situational awareness
- Communication (turns, count, deviation)
- Learnability
- User Reliance
- User Acceptance

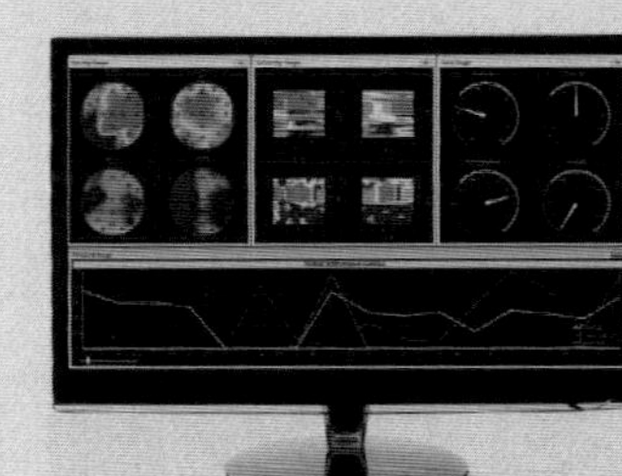

Defining Usability Metric Methods and Processes

INTUITIVE®

- While defined parameters help in writing the requirements and specifications, how these metrics are measured must be defined.
- Without defining the data capture and analysis procedures, the evaluation methods can be manipulated to get desired results.

The following considerations include:

- What is being measured?
  - o General or specific task workflow interaction with a UI?
- How is it being measured?
  - o User study, surveys, observations, etc.?
- Who are the evaluation participants?
  - o Developers, subject matter experts, end-users, etc.?

UNCLASSIFIED, DISTRIBUTION A

9

Specialized Users

INTUITIVE®

- DoD systems are generally designed for trained specialized users.
- When errors are encountered due to poor usability, the easiest solution and response is to train the user on how to avoid the error instead for fixing the issue.
  - Teaches non-intuitive solutions.
  - This burdens the operator with extra cognitive recall instead of using intuitive recognition.

UNCLASSIFIED, DISTRIBUTION A

10

Lack of Usability Specialists

INTUITIVE

- There are three main components to a system:
  - The Software
  - The Hardware
  - The User
- With project roles dedicated to software and hardware, it makes sense to have someone dedicated to the user and the usability of the system.
- Without formal usability training, non-usability specialist efforts may produce unintentional misleading or incorrect results.
- People feel like they are UX experts and have strong opinions on how the user will react.

"We've been doing this way for years without a so called 'user experience expert'. Plus, everyone on the team already understands what the user wants and needs."

Have a Dedicated & Trained Usability Specialist!

Non-Revenue Generation

INTUITIVE

Most government and DoD systems are not developed for the purpose of generating revenue. They are designed to perform or accomplish a specific and complex task.

In the private sector:

- Products that provide a function are often designed to sell volumes of that product.
  - o Examples: Microsoft Office, video games
- Products may be developed as a method to sell items.
  - o Example: Retail websites
- Product sustainment may rely on selling advertisements.
  - o Examples: Facebook, news outlets

When revenue is involved, additional KPIs can be utilized including:

- Sales generation
- Order amounts
- Number of pages view
- Total number of selections/clicks
- Dwell time

Conclusion

INTUITIVE

There are several factors contributing to usability shortcomings with DoD government developed and funded systems including: funding model, lack of competition, requirements and specifications, usability metrics, specialized users, lack of usability specialists, non-revenue generation.

- Upfront funding model and lack of competition during the development phases are practices that are difficult to change.
- Having UX Professionals involved early, and if appropriate usability SoW requirements are generated, system usability should greatly improve.
- Implement DoDI 5000.95 - HSI Defense Acquisition
- Even if usability requirements and specifications do not stipulate any usability metrics, outlining good usability practices, procedures, and accountability will greatly advance the system.
    - Requiring usability evaluations with end-users throughout the design and development phases will identify usability shortcomings early and greatly improve the overall system and end-user acceptance and satisfaction.

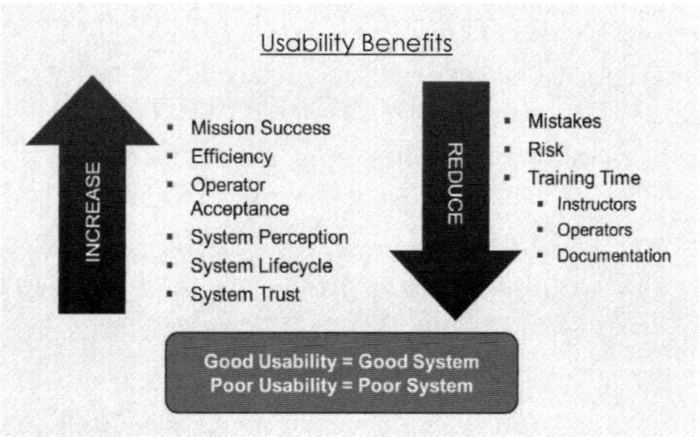

UNCLASSIFIED, DISTRIBUTION A

13

Questions?

INTUITIVE

Jerry.Burpee@irtc-hq.com
Jerry.B.Burpee.ctr@Army.Mil
256-656-0657

Jerry Burpee

The Usability Research and Evaluation Sr Principal Engineer for Intuitive Research and Technology Corporation in Huntsville, Alabama.

He is a U.S. Navy veteran and is the Human-Machine Interaction (HMI) subject matter expert for many U.S. Army's Air and Missile Defense (AMD) and Unmanned Air Vehicle (UAV) programs.

For the past 15 years he has led a dedicated HMI team conducting numerous military system and product usability research and evaluations with Army and Air Force Warfighters.

He has designed several usability evaluation laboratories, including the HMI Mobile laboratory, a trailer built to go directly to the Warfighter to perform usability evaluations.

Chair: Human Factors Ergonomic Society (HFES) Usability & System Evaluation (USE) Technical Group

Book: Burpee, Tenhundfeld 'Usability Evaluations & Usability Laboratories'

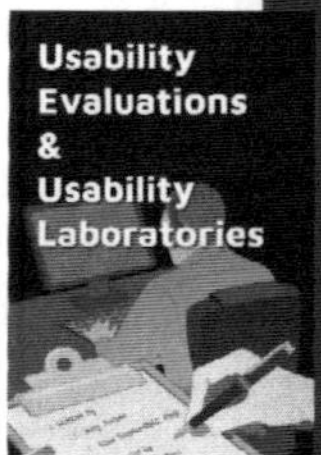

Thank You!

UNCLASSIFIED, DISTRIBUTION A

14

U.S. ARMY

DEVCOM AVIATION & MISSILE CENTER

U.S. ARMY COMBAT CAPABILITIES DEVELOPMENT COMMAND AVIATION & MISSILE CENTER

Rotor Blade Position Simulation from Dynamic Maneuvers for Launched Effects | **22-23 JAN 2025**

SRD AEROMECHANICS | DR. CHRISTIAN BRACKBILL, MODELING AND SIMULATION TEAM

SRD AEROMECHANICS | DR. JUSTIN MARTIN, MODELING AND SIMULATION TEAM

ROTOR BLADES AND LAUNCHED EFFECTS

DEVCOM AVIATION & MISSILE CENTER

- **Airworthiness:** DEVCOM-AvMC Systems Readiness Directorate **(SRD)** is the delegated Airworthiness Authority for the Army
- **Launch/Separation Envelope:** Certain types of launch systems can potentially impact the rotor system during dynamic maneuvers
- **SRD Mission Equipment Division:** Evaluates safe launch / safe separation for stores and launched effects
- **SRD Aeromechanics Division:** Evaluates (simulates) aircraft dynamic maneuvers to determine attitudes, rates and blade positions
- **Presenters:** Dr. Christian Brackbill, Dr. Justin Martin SRD Aeromechanics, Comprehensive Modeling Team

An ALE system is launched from a UH-60L Black Hawk helicopter, as testing commences during Project Convergence in September 2020. (Spc. Javion Siders/U.S. Army) (defensenews.com)

AGENDA

Video clipped from: *https://www.youtube.com/watch?v=eOeqzH7ZOgE*

- **Background – launch / separation envelope**
  - Types of launch / fires / separation
  - Types of maneuvers applicable to this analysis
- Big Picture – Airworthiness Process
  - SRD Mission Equipment responsibilities
  - SRD Aeromechanics piece of the puzzle – rotor blade position during maneuvers
- Modeling and Simulation
  - FLIGHTLAB, Virtual Pilot, RAAS
  - Sample Maneuver Analysis
- Additional Applications, Future Work

TYPES OF LAUNCH/FIRES/STORE SEPARATION

When do we care about Dynamic Maneuvers and rotor blade envelope?

- Missile Firing
- Countermeasures
- Wing-born ALE Launch
  - ***Angle and speed of launch, combined with transient rotor blade flapping, can bring the launch cone/envelope close to the rotor blade envelope***

When are Dynamic Maneuver *less* important?

- Gunfire
  - Typically use worst-case static blade droop and install range of motion limiters on gun travel
  - Projectiles clear rotor blade envelope very quickly
- Stores Jettison
  - Typically concerned with rotor aerodynamic influence on the Stores item
  - Analysis focuses on the safe motion of the separated Store away from the aircraft during relatively benign flight conditions

TYPES OF MANEUVERS USED FOR LAUNCHES

What types of Dynamic Maneuvers do we focus on first?

- Level Flight (baseline)
  - Compare trim blade position to dynamic maneuvers
- Flare (nose-up)
  - Countermeasures, wing-born ALE
- Bump / dive / pull-out / pushover
  - Ground targets

What additional "Engineering Maneuvers" do we consider for full qualification of a launch system?

- Side and rear flight
- Climb and descent
- Transition, sideslip, collective drops
- Roll, Lateral, Turns

AGENDA

- Background – launch / separation envelope
  - Types of launch / fires / separation
  - Types of maneuvers applicable to this analysis
- **Big Picture – Airworthiness Process**
  - SRD Mission Equipment responsibilities
  - SRD Aeromechanics piece of the puzzle – rotor blade position during maneuvers
- Modeling and Simulation
  - FLIGHTLAB, Virtual Pilot, RAAS
  - Sample Maneuver Analysis
- Additional Applications, Future Work

AIRWORTHINESS PROCESS

- **Focus of this briefing: Aeromechanics Process**
  - Dynamic maneuvers used for launch
  - Modeling and Simulation
  - Aeromechanics analysis of maneuver results

- **Overall Airworthiness Process: SRD Mission Equipment Division**
  - Dynamic maneuver simulations are performed by SRD Aeromechanics
  - Vehicle state data and rotor blade position data are analyzed by ***SRD Aeromechanics*** and passed to ***SRD Mission Equipment*** for Safe Separation / Launch Envelope analysis
  - Launch Envelope analysis may result in:
    - Published maneuver limits for safe launch (e.g. speed, attitude, rate or load factor limits)
    - Software-enabled launch limits (when desired / available)

SAMPLE REQUIREMENT

- MIL-STD-1289D
- 4.3.2.8 Propeller and rotor disk clearance
  - b. Rockets/missiles: The **clearance during launch** for guided and unguided rockets and missiles **shall be a five-degree half angle cone measured from the trajectory of the outermost surface of the ordnance to the worst-case rotor plane** or aircraft structure. The definition of a five-degree half-angle cone is depicted on figure 1. Clearance shall be sufficient to preclude induced damage from spent cases or any loose items under a worst case release condition. **(The worst case rotor plane cannot always be located accurately in a static condition and should also be measured in a dynamic situation.)**

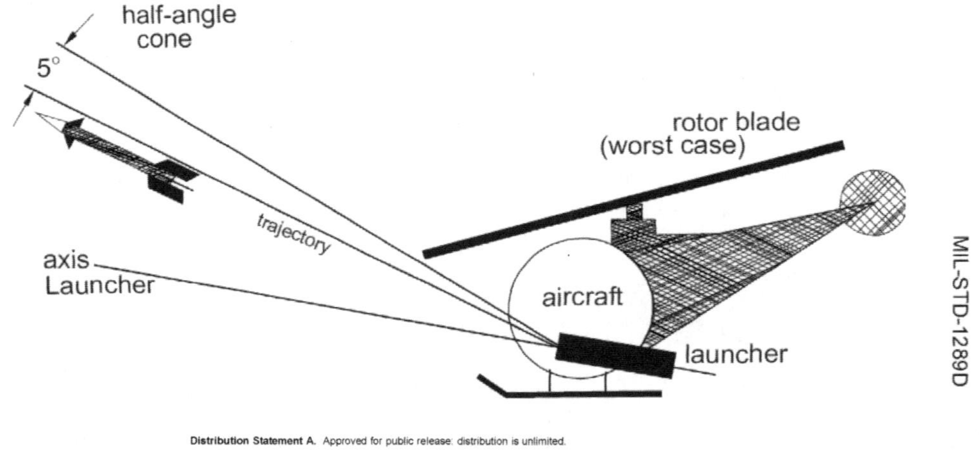

SAMPLE SRD MISSION EQUIPMENT ENVELOPE ANALYSIS

Mission Equipment Division Analysis (animation)

- Considers worst-case blade position
- Does NOT consider rotor azimuth angle (most conservative case)
- Can also consider x-y-z blade position of all 4 blades (or blade flap and rotor azimuth) to determine tip path plane angle, using data from Aeromechanics simulations
- Dive angle and low load factor contribute to significant blade dynamic flapping response

BLADE POSITION – IT'S ALL RELATIVE!

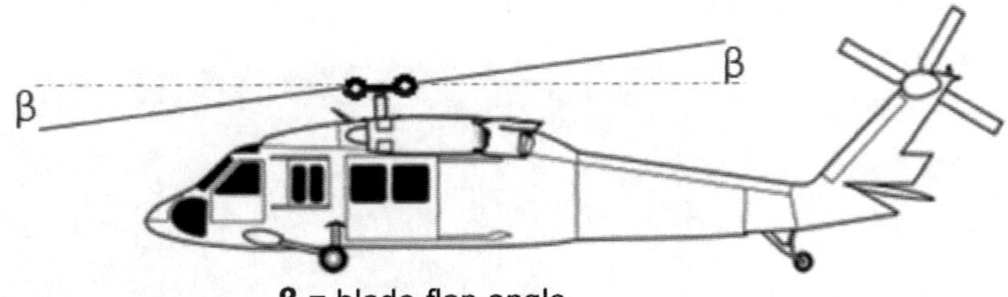

ß = blade flap angle

- Need to know
  - Aircraft pitch attitude and level reference
  - Mast tilt relative to reference
  - Pre-cone angle and coning angle
  - Cyclic blade flapping
  - Rotor Azimuth Position – if using flap angle; otherwise, you also need:
  - Blade tip X-Y-Z position
- FLIGHTLAB simulation software provides:
  - Aircraft attitudes, rates, etc
  - Blade flap angle ***relative to*** the blade flap hinge (for each blade, after precone is applied)
  - Azimuth angle of each blade ***relative to*** the tail boom
  - Rotor blade tip X-Y-Z position ***relative to*** the rotor hub plane

AGENDA

- Background – launch / separation envelope
  - Types of launch / fires / separation
  - Types of maneuvers applicable to this analysis
- Big Picture – Airworthiness Process
  - SRD Mission Equipment responsibilities
  - SRD Aeromechanics piece of the puzzle – rotor blade position during maneuvers
- **Modeling and Simulation**
  - FLIGHTLAB, Virtual Pilot, RAAS
  - Sample Maneuver Analysis
- Additional Applications, Future Work

MODELING AND SIMULATION PROCESS

- Model development and V&V
  - FLIGHTLAB
  - Autopilot (SRD "Virtual Pilot")
  - RAAS
- Maneuver analysis – SRD process
  - Maneuver Criteria
  - Numerical and Autopilot Trim
  - Autopilot guided (scripted) maneuver
  - Autopilot guided (scripted) recovery to trim
  - Output data processing and analysis
  - MATLAB scripts: SRD RAAS plotting and analysis tools
  - Customer Data Product: plots or data sets

https://en.m.wikipedia.org/wiki/File:Apache_Helicopter_Firing_Rockets_MOD_45154922.jpg

FLIGHTLAB AND VIRTUAL PILOT

- Generic Simulation Model
  - Single-Main-Rotor Helicopter (e.g. H60, H64)
  - FLIGHTLAB, RCAS, etc.
  - Fully nonlinear, low-to-medium fidelity
- V&V: SRD validates against available flight test data
- Autopilot control for simulation:
  - SRD "Virtual Pilot" Controller
  - Programmable in 4 axes or modes simultaneously *(Hold modes, maneuver targets, timing, etc.)*
  - Can follow control stick flight test time histories
  - Tuned and validated for each aircraft platform

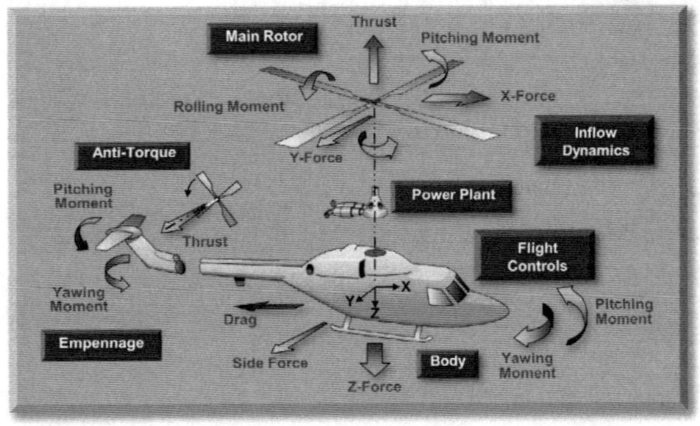

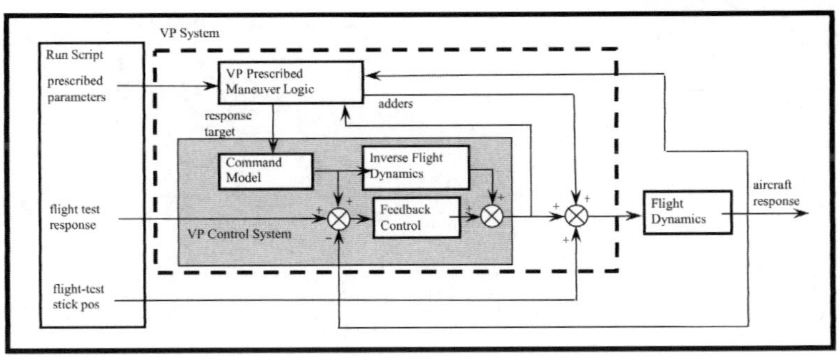

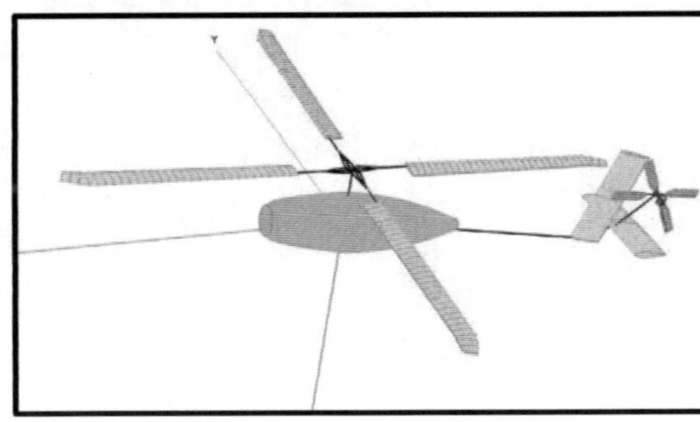

ROTORCRAFT AIRWORTHINESS ANALYSIS SUITE

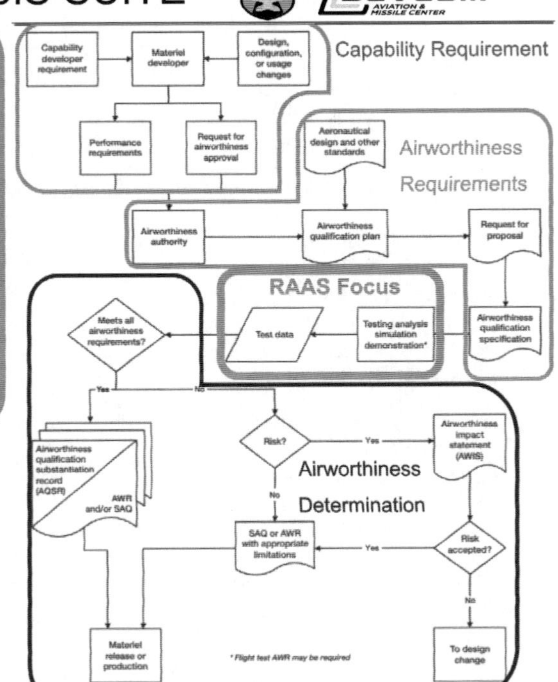

- RAAS is a framework that provides <u>Continuous Integration</u> of <u>data and models</u> to support and enable <u>initial</u> and <u>continued</u> <u>airworthiness</u>
- RAAS Mission:
  - Accelerate Airworthiness Analysis
- RAAS can link externally to: MBSE / TTS / Customer Databases

MANEUVER CRITERIA: INPUTS AND CONSTRAINTS

- Desired maneuver inputs and constraints come from discussions with:
  - Pilots
  - Flight test engineers
  - Developers/OEM
  - Other stakeholders

- Example: flare (pitch-up)
  - Establish aircraft in trim flight at target airspeed
  - Initiate flare (pitch-up) to target vehicle pitch attitude
    - Target attitude acquired in 1-2 seconds, max
  - Capture the target pitch attitude
    - Hold for 2-3 seconds
    - Some airspeed loss is acceptable (or not)
    - Some altitude gain/loss is acceptable (or not)
  - Recover to
    - Original trim condition, OR
    - Some other trim condition

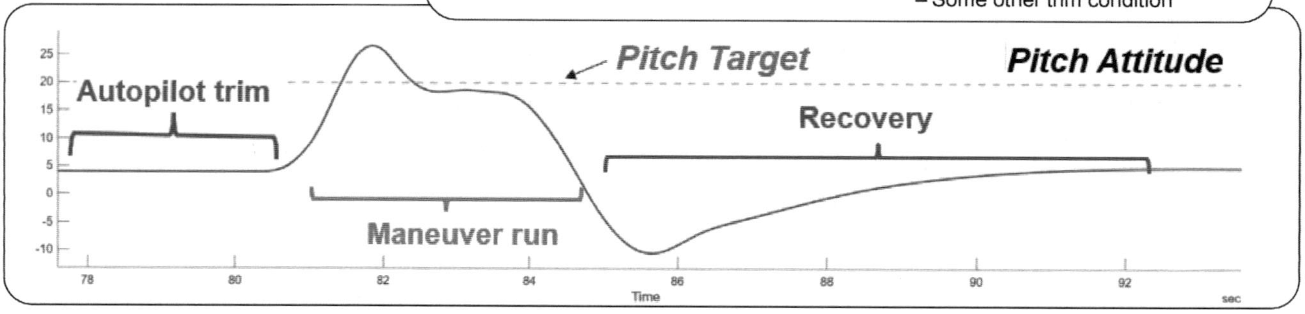

MANEUVER RUN PROCEDURE

- Simulation Time History
  - Includes more than just the desired maneuver
- Process
  - Trim numerically
  - Engage Virtual Pilot (autopilot)
  - Obtain autopilot trim at the original trim condition
  - Perform the maneuver
  - Recover to desired condition
- Simulation run time:
  - Hold/Wait times ensure that transients die out before and after the maneuver
  - Model runs **faster** than real-time so longer maneuvers are not a problem

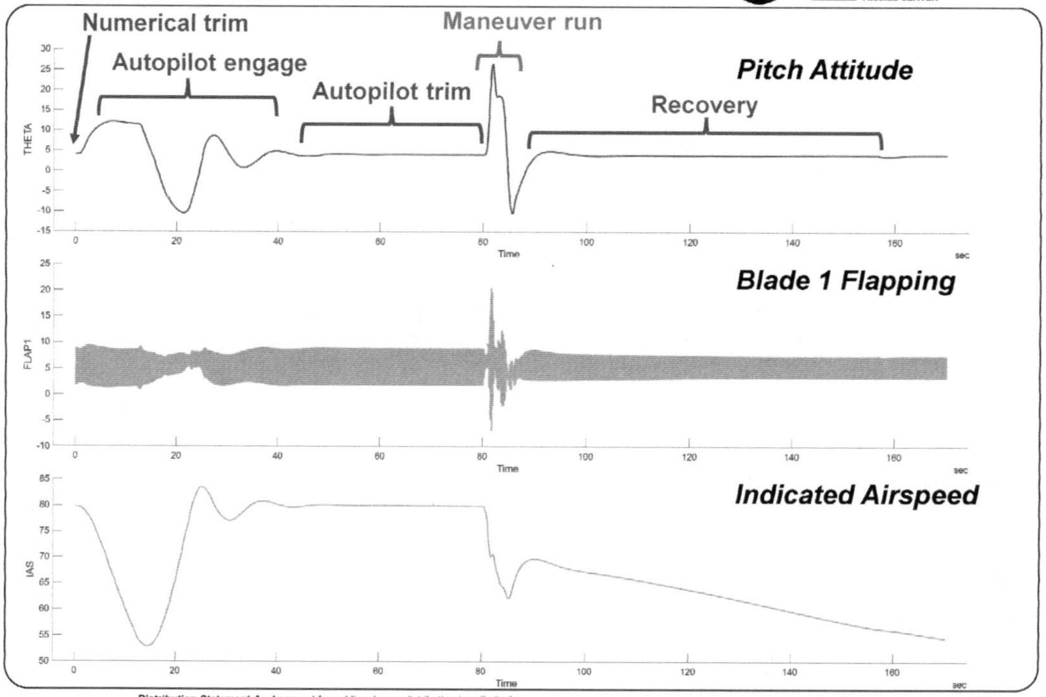

TOP LEVEL MANEUVER ANALYSIS

- **Sample Maneuver**
  - **Flare (pitch up from trim)**
- Blade Flapping
  - **Does NOT show rotor azimuth information**
  - Rotor Azimuth can be critical for envelope calculations
- **NOTE:** Flap Limit is **"Notional"** because each aircraft and launch system combination can have a different flap limit

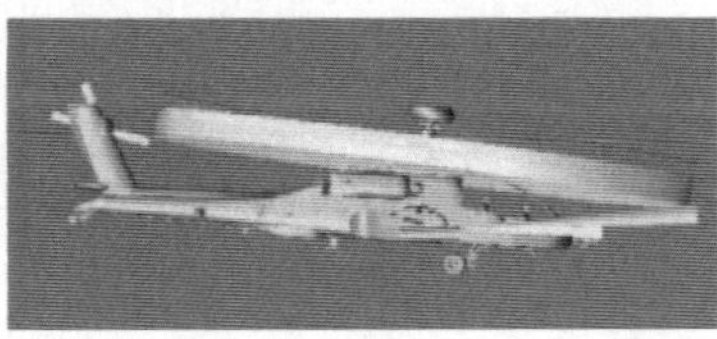

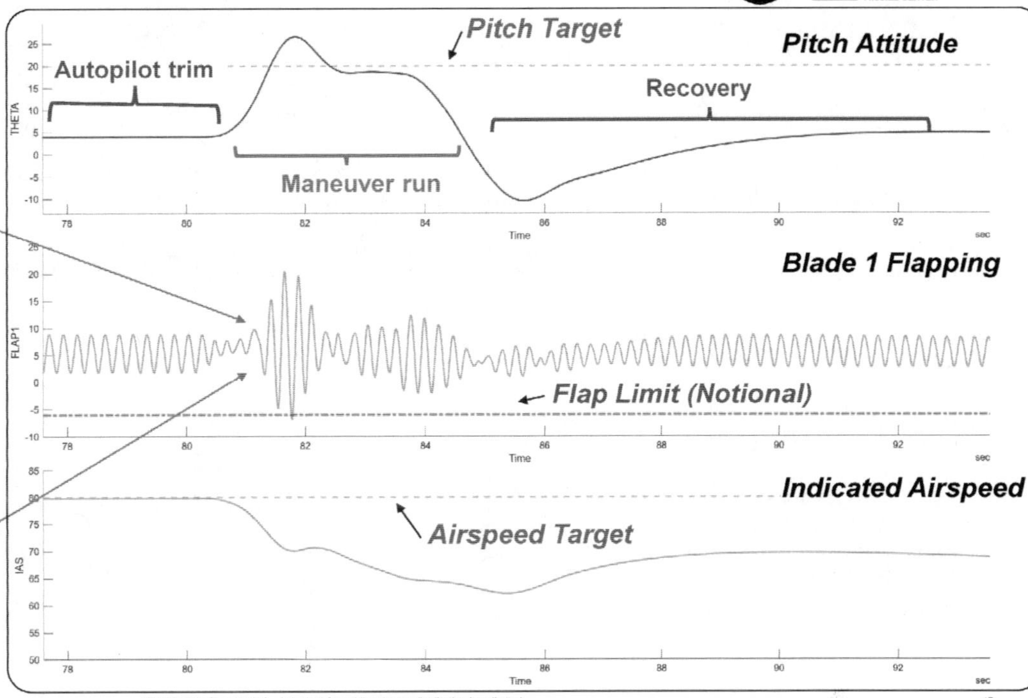

BLADE FLAPPING WITH ROTOR AZIMUTH INPUT

- **Sample Maneuver**
  - **Flare (pitch up from trim)**
- Blade Flapping
  - **Post-process flapping with rotor azimuth data**
  - Plot time-history of blade flapping over the aircraft nose
- **NOTE: "Aircraft Nose"** flapping refers to the observed blade flap angle directly in front of the aircraft.
  - Combines all four blade flap angles and the rotor azimuth angle
- What causes the large negative flapping in this case?
  - It is NOT the dynamic pitch-up (see next chart)

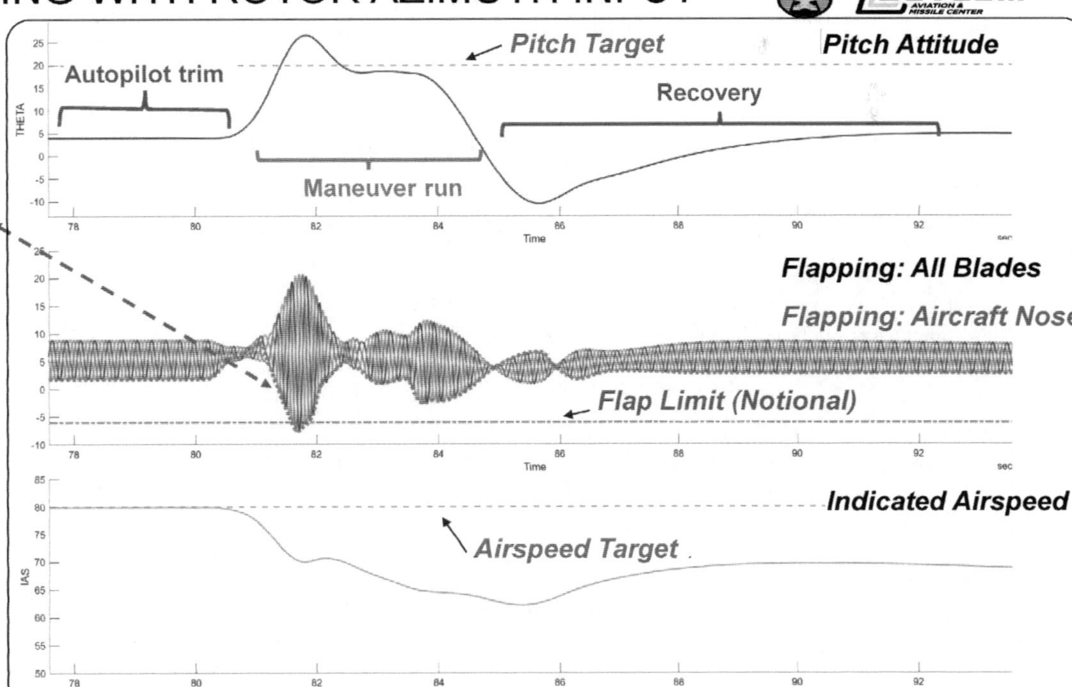

BLADE FLAPPING WITH FLIGHT CONTROL INPUT

DEVCOM AVIATION & MISSILE CENTER

- **Sample Maneuver**
  - **Flare (pitch up from trim)**
- Blade Flapping
  - Negative blade flapping at the aircraft nose
  - Results from large, rapid forward cyclic application by the autopilot
- Autopilot
  - Scripted and tuned
  - Goal was ~1 second rise in pitch attitude to the target followed by ~2-3 seconds at the target pitch
  - Results in overshoot based on current tuning parameters
  - Overshoot leads to large, rapid (unrealistic) forward cyclic application

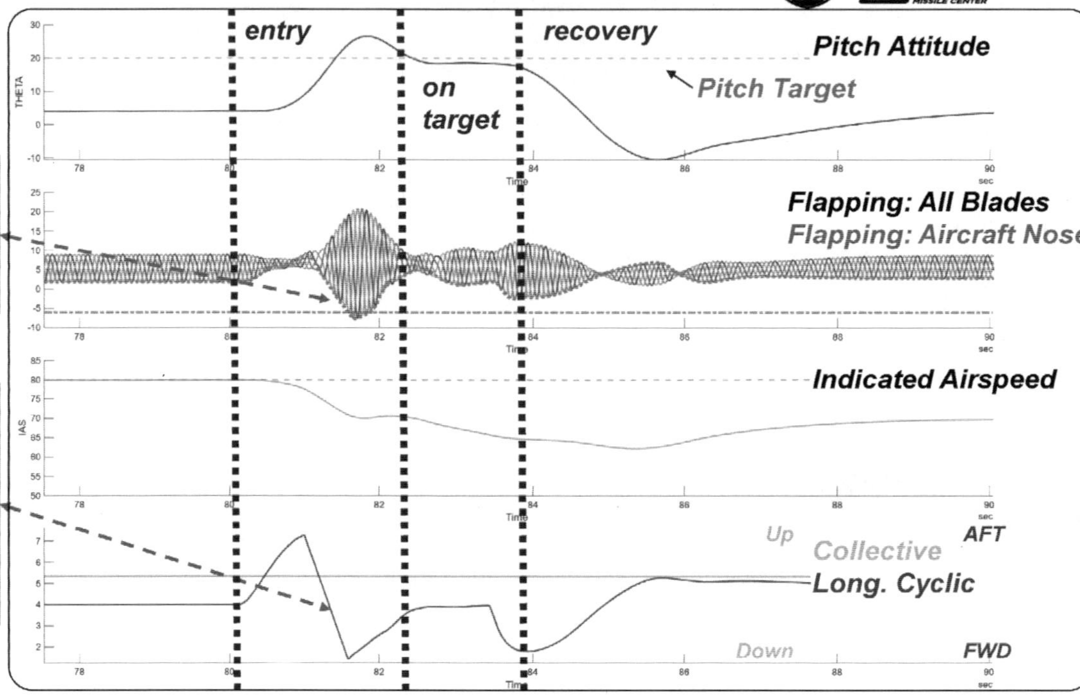

A WORD ABOUT PROGRAMMABLE AUTOPILOTS

DEVCOM AVIATION & MISSILE CENTER

- SRD Virtual Pilot
  - Tuned and validated ***for maneuvers available in the flight test record***
  - Programmable, up to a point
  - Capable of following flight test control inputs ***if they are available***
- For some of these maneuvers:
  - We currently don't have test records or control input time histories
  - Thus: maneuvers are programmed or scripted in the autopilot
- What could be changed in the scripting?
  - Controls:
    - Collective bump to maintain altitude (modifies blade coning – flap margin)
    - Longitudinal Cyclic: more rapid but smaller inputs, longer holds (fast, small motion; hold until approaching pitch target, reverse to capture target)
- If we had the flight test data:
  - Retune the autopilot to better match data, if available
  - Use flight test controls directly to improve autopilot behavior

COMPLETED "CUSTOMER" PLOT (SAMPLE)

(BUILT WITH RAAS GUI OR RAAS BATCH SCRIPTS)

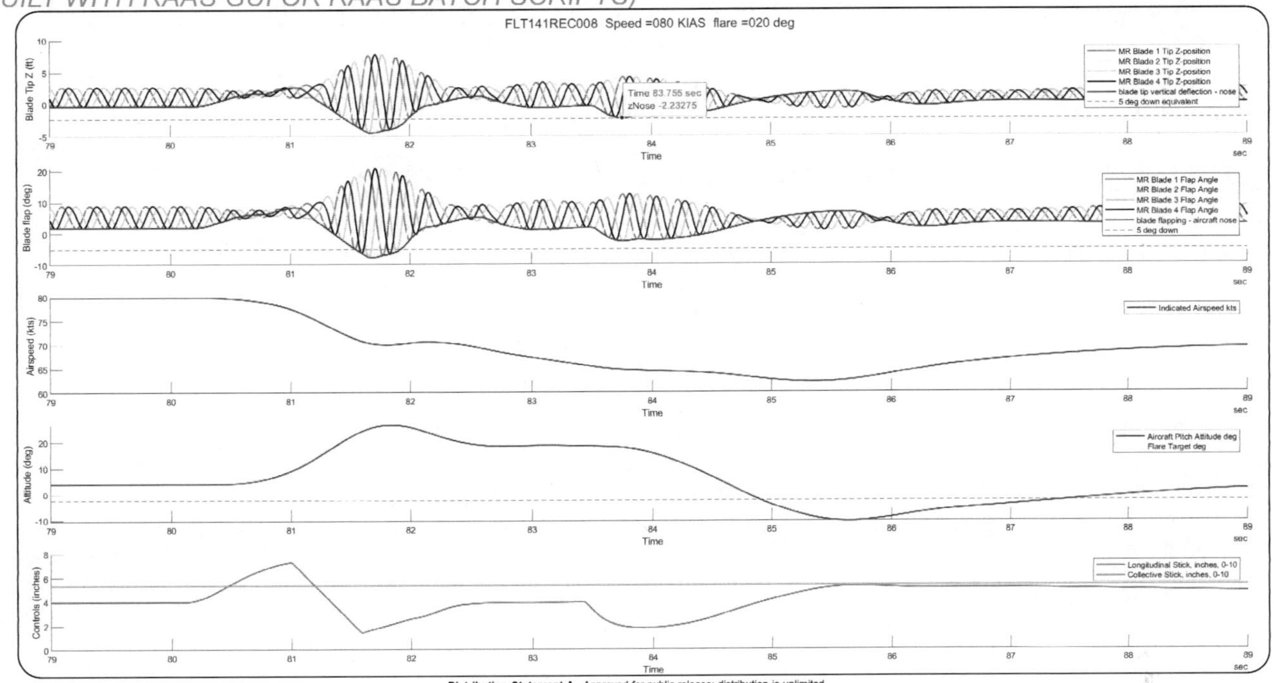

RAAS FLIGHT DATA VIEWER

- Basic plot building and analysis using the RAAS GUI
  - Flight Data Viewer (FDV)

- *Note:*
  - *Video is ~5 minutes in real time*
  - *Includes building and customizing one plot in RAAS*
  - *Final customer plots are then scripted and batch-processed offline*

SAMPLE USE CASE – FLIGHTLAB AND RAAS

- Use SRD RAAS (MATLAB) tools *(Batch scripts for analysis of large data sets)*
  - Build a database of simulation results
  - Batch-process all cases
  - Rapidly conduct parametric studies
- Maneuvers run faster than real-time
- Batch-convert output files to RAAS database format
- RAAS Interactive plotting & analysis
- RAAS batch-post-processing for reporting
- Plots show both blade flapping and blade tip Z position, confirming they show the same trend.

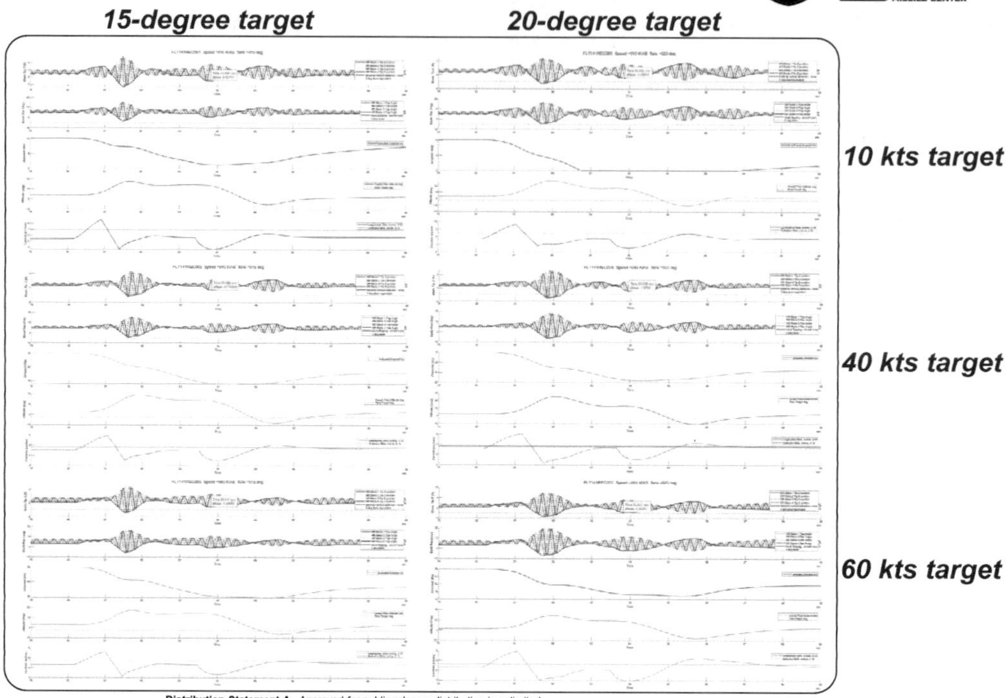

AGENDA

- Background – launch / separation envelope
  - Types of launch / fires / separation
  - Types of maneuvers applicable to this analysis
- Big Picture – Airworthiness Process
  - SRD Mission Equipment responsibilities
  - SRD Aeromechanics piece of the puzzle – rotor blade position during maneuvers
- Modeling and Simulation
  - FLIGHTLAB, Virtual Pilot, RAAS
  - Sample Maneuver Analysis
- **Additional Applications, Future Work**

AGGRESSIVE LOW SPEED PUSHOVER

- **Sample: Pushover Case**
- Maneuver is on target from ~65-66 seconds
- Worst case blade flapping occurs on target, but is due to aft cyclic pull that initiates the recovery
- Azimuth data was not included with this output data set, but this large negative flapping should be occurring over the tail boom due to the aft cyclic pull. (Verify with blade x-y-z position data.)

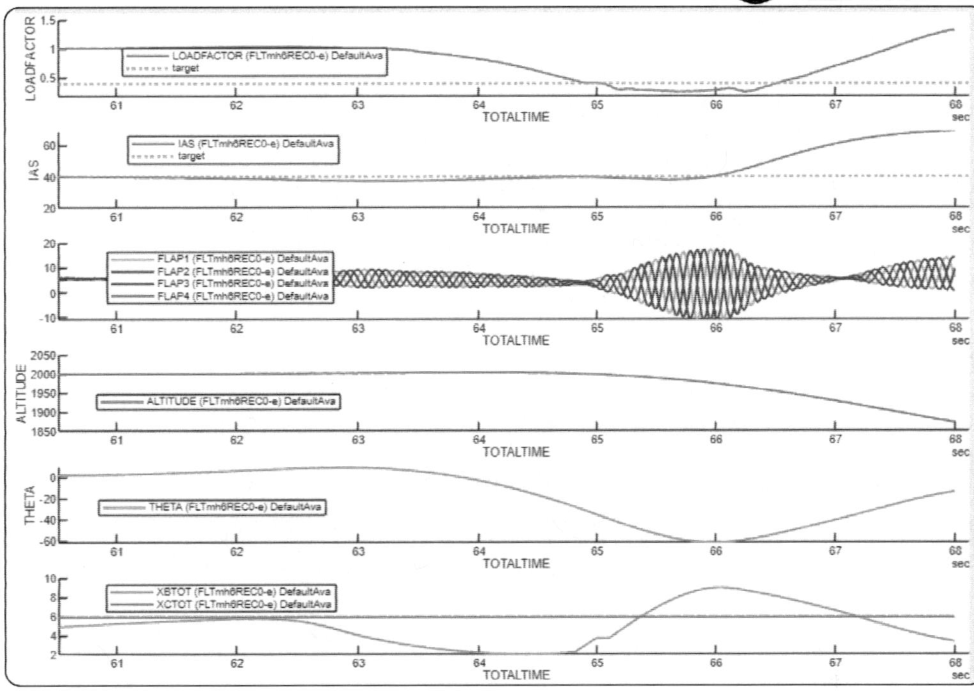

AGGRESSIVE MODERATE SPEED DIVE

- **Sample: Dive Maneuver**
- Note that autopilot recovery is not well modeled
- Target maneuver (~64-67 seconds) is well modeled
- There is no significant ***negative*** blade flapping, even during the overly aggressive autopilot recovery

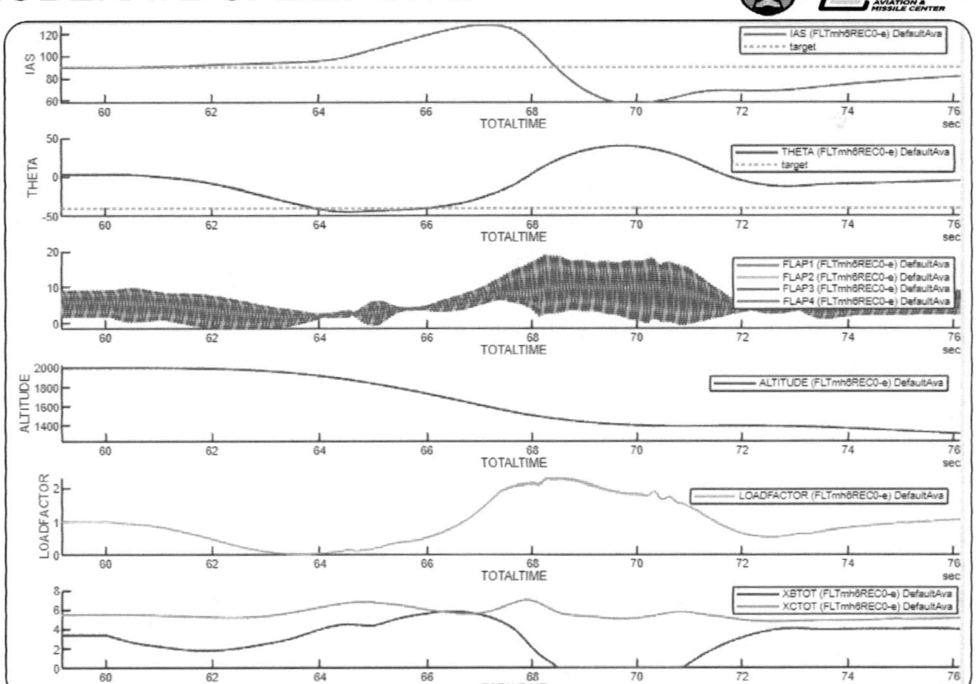

FUTURE WORK

- **Automated Database from Simulation results**
  - FLIGHTLAB and RCAS batch simulation runs will automatically be sent to a database
  - RAAS tools will enable database searches on Record/Event, Maneuver type, sensor data, Aircraft Type/Platform, etc.
  - **Allows for automated data transfer between Aeromechanics and Mission Equipment** (or integration of Mission Equipment simulations with RAAS infrastructure)
  - Allows for tagging/tracking/time-stamping simulation results (instead of manually labeled share-drive storage)

- Piloted Simulation Results Database
  - SRD Piloted Simulation time histories can be tagged and stored in a similar database
  - Same process we use today for the RAAS flight test database

- Modernized Virtual Pilot (Autopilot)
  - Additional flight test time histories for these types of maneuvers will allow for model refinements and control-following simualtions

Distribution Statement A. Approved for public release: distribution is unlimited.

Distribution Statement A. Approved for public release: distribution is unlimited.

ACKNOWLEDGEMENTS

- Tuan Pham, SRD Mission Equipment Division
- Wesley Viall, SRD Structures and Materials Division

Distribution Statement A. Approved for public release: distribution is unlimited.

DEVCOM AVIATION & MISSILE CENTER

THANK YOU.

12 February 2025

29

U.S. ARMY

Vertical Flight Society
2700 Prosperity Ave, Suite 275
Fairfax, VA 22031
USA

ISBN 979-8-3313-2655-5

9 798331 326555

CURRAN ASSOCIATES INC.
proceedings.com